CW01188024

German Fighter Ace Werner Mölders

German Fighter Ace
Werner Mölders

An Illustrated Biography

Ernst Obermaier & Werner Held

Schiffer Military History
Atglen, PA

Book translation by Christine Wisowaty

Book Design by Ian Robertson.

Copyright © 2006 by Schiffer Publishing.
Library of Congress Control Number: 200692694

All rights reserved. No part of this work may be reproduced or used in any forms or by any means – graphic, electronic or mechanical, including photocopying or information storage and retrieval systems – without written permission from the copyright holder.

Printed in China.
ISBN: 0-7643-2526-4

This book was originally published in German under the title
Jagdflieger Oberst Werner Mölders by Motorbuch Verlag

We are interested in hearing from authors with book ideas on related topics.

| Published by Schiffer Publishing Ltd.
4880 Lower Valley Road
Atglen, PA 19310
Phone: (610) 593-1777
FAX: (610) 593-2002
E-mail: Info@schifferbooks.com.
Visit our web site at: www.schifferbooks.com
Please write for a free catalog.
This book may be purchased from the publisher.
Please include $3.95 postage.
Try your bookstore first. | In Europe, Schiffer books are distributed by:
Bushwood Books
6 Marksbury Avenue
Kew Gardens
Surrey TW9 4JF, England
Phone: 44 (0) 20 8392-8585
FAX: 44 (0) 20 8392-9876
E-mail: Info@bushwoodbooks.co.uk.
Visit our website at: www.bushwoodbooks.co.uk
Free postage in the UK. Europe: air mail at cost.
Try your bookstore first. |

Contents

Foreword 7
Oberst Werner Mölders - The Life of a German Fighter Pilot 9
A Timeline of Werner Mölders' Life 31
The Mention of Werner Mölders and his *Geschwader* in the *Wehrmachtbericht*
    OKW-Bericht), Special Announcements and Press Reports. 37
List of Werner Mölders' Air Victories 40
Childhood and School Days 44
Military and Aviation Training 49
As a Fighter Pilot in the new German *Luftwaffe* 71
With the "Legion Condor" in Spain 78
Missions on the Western Front 93
Missions on the Channel 109
Missions on the Eastern Front 157
Inspekteur of the Fighter Pilots 176
Farewell to Werner Mölders 191
Unforgotten - *Oberst* Werner Mölders 209
Sources 230

Bf 109 - *Vierfinger* - *Schwarm*.
Mölders played the leading role in the development of this combat formation, also called the "Mölders-Formation" by Englanders.

Foreword

The most impressive experience I had during my time on the Front in World War II was meeting *Oberst* Werner Mölders.

I was an amateur pilot in 1939 when the war broke out, and arrived at the Front to the *Jagdgeschwader* 51 in the summer of 1940 after fighter pilot training. At this point in time a *Geschwader* was like any other, and no one could have guessed that *Jagdgeschwader* 51 would be one of the most famous air units on both sides.

In the battle for England, the *Geschwader* was stationed in Northern France near the coast. I was on duty as a pilot in the I. *Gruppe*, and was ordered to report to the *Kommodore* for an escort assignment. Surely I had heard much about him, saw him a couple of times, and was proud, like my comrades, that our *Kommodore* had the most air victories, but I still did not have personal contact with him. I could not wait to meet him.

In a conversation I got to know the *Offizier* who had so much charisma and, irrespective of my rank, spoke simply as a comrade, without ever seeming jovial or being overbearing; without hesitation I said yes to his suggestion at the end our of conversation that I accompany him on the mission.

In the following months we lived under one roof, I flew together with him, experienced the routine of his missions, his way of dealing with critical situations, his command of the *Geschwader*, and his circle of comrades, and was witness to many open conversations with his superiors. This man, who in his outward appearance seemed very correct and often serious, who could truly be happy with us, who was open and brave, became a model for us fighter pilots.

Maybe it was these characteristics, along with his honesty and practicality that were the reason that we, his technicians, pilots, soldiers, officers, enlisted men, and simply everyone who he was responsible for, convincingly spoke of him as "*Vati* Mölders."

From my perspective and experience, few men have the talent to evoke trust, to be convincing while remaining a friend and, more so, to unconditionally be there and ready for his comrades in critical situations. This is what I learned of the *Kommodore* in the short time of our meeting.

A Model? For the men who see these human characteristics as positive, yes!

Erwin Fleig

Oberst Werner Mölders
The Life of a German Fighter Pilot

On November 23, 1941, the then young author of this book heard the following report on the German radio:

> As the *Oberkommando* of the *Wehrmacht* announced, the *Inspekteur* of the German fighter pilots, *Oberst* Mölders, was killed in a plane crash. The crash occurred in a plane of which he was not at the controls.

I remember this announcement and my reaction so well as if it were just today. I heard the announcer's words, which were unbelievable; I felt as if paralyzed, broke out into tears, and for days was hardly responsive or consoled. "My" idol and model as a pilot, soldier, and man was dead! A man, who was not only admired so much by me, but also by hundreds of thousands of young people my age, even from the millions of Germans during those years.

Whoever reads this today and did not experience the years between 1938 and 1945 for themselves will probably astonishingly ask, how was that possible? Who was this man that was admired by so many? What distinguished him so much from the others, that the masses paid attention to the fate of a single person during a time in which each was concerned enough with their own fate?

The author wants to try to give answers to these questions with this book. The attempt—it can be no more—to pay tribute to Werner Mölders with this illustrated book, requires such a number of starting points, that it is not easy to arrange them correctly and to describe to the reader so that s/he understands how no one has forgotten about this soldier even today, and that—until now once in the history of the German armed forces—his name is represented in both the Marines and *Luftwaffe* of the *Bundeswehr*!

So who was *Oberst* Werner Mölders, who every German then knew as the most successful soldier of the *Luftwaffe* from the papers, magazines, and news reels: his wiry figure, his tanned face, the serious, honest, and warm-hearted look in his eyes?

Mölders was born on March 18, 1913, in Gelsenkirchen as the son of a secondary school teacher. He hardly knew his father, because very early on he fell victim to World War I. He fell on March 2, 1915, as *Leutnant* of the Reserves in *Königsinfanterie-Regiment* No. 145 in L'Argonne (France). The young widow moved with her children, three boys and one girl, into her parents' house in Brandenburg a.d. Havel after his death.

There Mölders spent his childhood and youth. He grew up in this amazing *märkisch* landscape into a person who loved and felt in tune with nature. His Uncle Paul, who lived in the country in close proximity to Brandenburg and who he especially liked, brought him closer to the world of animals. His entire life he felt a special relationship to nature, to the wild, and to the noble art of hunting.

In Brandenburg a.d. Havel he attended the "Saldria-*Gymnasium*" and discovered his love for water sports as a schoolboy. On beautiful summer days he laid down with his comrades next to the Havel and achieved as a student rower the *RV. "Saldria-Brandenburg,"* and later regatta success in the "Brandenburger Ruderclub." Mölders felt well on the water, and he passionately put everything that he had into this sport. When he came to the "*Brandenburger Ruderclub*" he soon led a division of the *Jung-Ruderer*.

Next to water sport clubs Mölders also belonged to the *Bund Neudeutschland* in the Catholic youth movement. Together with his younger brother, Victor, whom he was always especially close to, he felt the urge to get out into the countryside in this community, to camp with like-minded comrades in his free time. Because of his honest nature, and his friendly and exemplary sporting manner, recognized and loved by all, he became group leader of the youth organization.

Raised religiously from childhood on and in the young community led by Kaplan Klawitter—his lifelong minister and friend—his faith in God grew, and he remained loyal to his beliefs even as *Offizier* and soldier until his early death. His great human virtues surely resulted from his views on God and the Catholic Church.

As a boy, Mölders always wished to become a soldier. After he completed his school-leaving exam, he volunteered for the *Reichswehr* at 17.5 years old. From 60 *Offiziersanwärter* of his age group that applied for the "*Hunderttausend-Mann-Heer*" with the *Infanterie-Regiment* 2 in Allenstein/East Prussia, only three were accepted, one of which was Mölders. On April 1, 1931, the military career of this German soldier, who today remains unforgotten, began.

After infantry basic training, Mölders came to the *Kriegsschule* Dresden in October 1932, and was transferred to the *Pionierschule* in München after the completion of his training in June 1933 as *Fähnrich*. Mölders always liked to remember his time in München as a *Fähnrich*; he seized every opportunity to indulge in his old love of water sports, on the Isar and Starnberger Lake. His love of mountains was also awoken here, and he went on many mountain hikes in the Bavarian Alps with his comrades.

The training of the budding *Pionieroffizer* was difficult and versatile, but it brought Mölders joy. During this time he volunteered for the new German *Luftwaffe* that officially did not exist then, but was "quietly" in the making. It was his childhood dream to be able to fly, and as the course participants were asked who would want to fly, he immediately answered.

The result of his pilot fitness test was a catastrophe! He could not bear the gyrostabilizer. "They sent me home sick and pale and unfit for flying," he noted in his diary. Despite that, Mölders did not give up. He wanted to reach his goal of flying. He was permitted to take the test once more and was accepted as "fit for certain duties."

After his following promotion to "Oberfähnrich" on February 1, 1934, he had finally become a pilot. Within the same month he came to pilot training at the *Deutsche Verkehrsfliegerschule* in Cottbus, in the Mark Brandenburg. *Leutnant* since March 1, 1934, he had initial difficulties in reaching his goal. Splitting headaches and dizziness tortured him during the first hours of training. But his willpower was stronger than physical weaknesses. When he flew at the controls, he overcame these difficulties and completed his training in Cottbus as the best in the course.

Subsequently, he came to further training at the *Kampffliegerschule* Tutow and to fighter pilot training in Schleiflheim, near München. During this time the formal taking over as *Leutnant* in the newly created and since March 1, 1935, existing German *Luftwaffe* followed, as well as the conferment of the *Flugzeugführerabzeichen* of the *Luftwaffe* on May 21, 1935.

From July 1935 until April 1936 Mölders flew with the *Fliegergruppe* Schwerin (I./162 "Immelmann"). During the occupation of the so-called "demilitarized Rheinland" on March 7, 1936, Mölders flew with his *Staffel* coming from Lippenstadt in Westfalen, over the Ruhrgebiet on the Rhine. He landed with his men in Düsseldorf, where they were enthusiastically greeted by the people.

During this "time in Schwerin" that, according to his own words, was considered the most beautiful and carefree time of his life, Mölders met *Frau* Luise Baldauf, whom he married during the war a few weeks before his tragic pilot's death.

On April 20, 1936, Hitler's birthday, many promotions followed in the new *Wehrmacht* and, as a result of April 1, 1936, Mölders was promoted to *Oberleutnant*. At the same time he was appointed to lead the *Jagdschulstaffel* in the II. *Gruppe* of the *Jagdgeschwader* 134 "Horst Wessel" that was stationed in Werl/Westfalen. The *Gruppe* was led by *Major* Theodor Osterkamp, one of the outstanding fighter pilots of World War I (32 air victories), awarded with the "Pour le Merite" medal, and known by all pilots of this time as "Uncle Theo." Mölders learned much from him, a passionate pilot and fighter who was also an exceptionally chivalrous and outstanding unit leader.

In order to become a fighter pilot, one needed not only courage and a good eye. The eye must also be trained for extensive observation of the air space: complete control of the plane and weapons, unconditional devotion to comrades, and chivalrous conduct in life as in combat were top priorities. Mölders could not have had a better *Kommandeur* and mentor than Theo Osterkamp. On March 15, 1937, Mölders became *Staffelkapitän*, and took over the 1. *Staffel* of the I./JG 334 in Wiesbaden.

In the summer of 1936 the Civil War broke out in Spain, where the German government sided with the Nationals under General Franco through the dispatch of the so-called "Legion Condor." In the German contingent there was a *Jagdgruppe* whose pilots were exchanged at certain intervals to give as many pilots as possible the opportunity to gain experience in war. *Kommandeur* Osterkamp registered Mölders for Spain, but he had to wait until Easter 1938 until it was his turn. In a letter to his brother, Hans, he wrote:

> "What had to be taken care of has been done. That was necessary in case I do not see you ever again. I have a strange inner conviction that nothing will happen to me. I have distributed my fortune and things; mother, Victor and you—each will receive their part. I did not forget my little godson Hartmut!"

Mölders was assigned to the 3. *Staffel* that was led by *Oberleutnant* Adolf Galland. Equipped with the inferior He 51, this *Staffel* was put into action as *Schlachtflieger*. Low-level attacks near the enemy's ground defense demanded a great deal of willpower and courage from the pilots. Under the leadership of "Capitano" Galland, the "bunny" Mölders completed his first missions until he took over the leadership of the *Staffel* on May 24, 1938, because the *Kämpe* Galland returned home.

The days of the He 51 were numbered; the risk had become too great, and again and again the excellent flak of the *Rotspanier* got its victims. Both *Schlachtfliegerstaffeln* of the "Legion" pulled out, and the arrival of the modern fighter planes, Messerschmitt Bf 109s, was eagerly awaited. At the beginning of July 1938 the time had finally come, and little by little the *Staffel* Mölders could be equipped. And already during the second mission the first successes were achieved.

Mölders took off with six planes. The first air victory is the most important for the fighter pilot. Mölders had to get a word in about his:

> "For a moment I had to really catch my breath...by Valencia I recognized four little dots from a distance...opponents. The moment longed for was there. I climbed five hundred meters higher and gave the sign for attack. The dots came closer...I recognized Curtisses, about forty to forty-five. There were six of us, all inexperienced, but they hadn't seen us yet. - Go towards them! - I approach, but shoot too soon out of nervousness - the guy turns around, came towards me - I see the muzzle of four MGs directed at me. - Damn it! I get scared: Firing is not that easy. There, a parachute, a Curtiss on fire - the first shootdown for my *Staffel*. Who could it have been? - Again around the crate and in between - A trail of smoke hangs in the air - two Curtisses coming towards me - I'm sweating like a pig - through - a Messerschmitt, that's crashing vertically - Hopefully it will steady itself - Thank God,

don't lose a plane today in the first air battle! 45:6, I reproach myself for attacking - further, merely attacking, I'm suddenly very calm, overlook the entire air space. Two Curtisses pull up - I dive under, come from below, the one behind notices me, flees, but I have the other one in sight - fifty meters - all four of my MGs hammer away - He rears up, tips over, I'm after him, once more the MG - He's smoking, he's on fire - first air victory - I am incredibly lucky - Where are me Mes?

The Curtisses are all forced away, I collect four of my planes and fly home. The planes fly over the area in a swerving motion (the signal for a shootdown). Behind me another plane flies in a swerving motion. It is *Leutnant* Lipper, who had the first shootdown for the *Staffel*. Below I see my mechanic jumping up and down for joy. My brave, first *Wart*, with the good German name Meier, can paint the first white line on the rudder assembly. *Leutnant* Oesau also shot down his first, so I could proudly report three air victories for our first air battle to my *Kommandeur*."

Mölders analyzed his first shootdown again and again. His excitement as he approached the enemy caused him to press the button too soon. During the next air battle he remained calm and came in closer. As *Staffelkapitän*, *Gruppenkommandeur*, and later as *Geschwaderkommodore* he saw it as his chief task to help the recruits with their shootdowns. It did not depend on bringing in single "Asse" with a high number of shootdowns, but his goal was to lead a large number of fighter pilots with average performance.

The first air victory on July 15, 1938, was not his last. With growing experience, cold-bloodedness, and calmness of nerves his number of shootdowns also increased. One must make sure to view these air battles in Spain as easy tasks. The Bf 109 was indeed seen as inferior to their *rotspanisch* opponents, but the Russian I-15 ("Curtiss") and I-16 ("Rata") were opponents to be taken seriously when they were flown by aggressive pilots. Mölders reported on air battles that lasted twenty minutes without a shootdown. The development of air combat in Spain took a great leap forward. One must imagine that during the beginning the fighter pilots moved with each other in a closed formation of 20, 30, or more planes. The German fighter pilots used the advantages that their modern Bf 109 planes offered, and developed a new kind of combat formation that was used by all war-waging air forces in the course of the next years; the so-called *Vierfinger-Schwarm*. In this loose formation the German fighters flew as *Rotten*, *Schwärme* (two *Rotten*), or *Staffeln* with a great amount of space in between and at various altitudes. A large air space could be covered, they did not have to desperately concern themselves with maintaining formation, the loose unit could be lost sight of, could see better, observe the air space behind the *Rotte* comrades, and eliminate surprise dangers. The small combat units *Rotte* and *Schwarm* had the freedom of deciding, and could flexibly adjust to the respective air situation. With the development of the *Vierfinger-Schwarm*, Werner Mölders played the leading roll. It is not without cause that, for example, in the Imperial War Museum in London this combat formation is called "Mölders Formation," and was also used by the Royal Air Force.

The number of his air victories grew steadily, and found recognition in his premature promotion to *Hauptmann* as a result from October 1, 1938. Even more noteworthy is that he won the title "*Vati Mölders*" in Spain, an expression of his strong father-like manner and thoughtfulness as superior and comrade. On December 5, 1938, the hour of parting arrived. With 14 acknowledged and 3 unconfirmed air victories, he returned to Germany as the most successful fighter pilot of the "Legion Condor." It was only natural that the leaders did not let him escape with his experiences, and they sent him to the *Reichsluftfahrtministerium* for three months, where he worked on the new fighter pilot rules with the *Inspekteur* of the fighter pilots. On March 15, 1939, he went back to the troop, and became *Staffelkapitän* of the 1. *Staffel*/JG 133, which soon became the 1./JG 53 "Pik As." On the occasion of his return from the "Legion Condor" to Germany, Werner Mölders was awarded with the *Deutsches Spanien-Kreuz in Gold mit Schwertern und Brillanten*.

When World War II broke out on September 2, 1939, Mölders' *Staffel* belonged to the units that took over border surveillance in the Mosel - Saar - Pfalz region. For him, the war did not begin very promisingly, because on 8 September he had to make a forced landing on a field due to engine trouble, and did a loop and back dive. It came to be that the first air victories of his *Staffel* were achieved without his assistance, which naturally riled him, particularly because his "little" brother Victor, a pilot in a *Zerstörergeschwader,* achieved an air victory with a Me 110 already in the first days of the military campaign against Poland, and received the EK II for it. But on September 20, 1939, it had come so far. During border surveillance in close proximity to where three countries meet, Mölders, with his *Schwarm*, surprised a *Staffel* of French Curtiss fighter pilots. After short turning-flight combat the Curtisses attacked by him crashed down, on fire, and the pilot saved himself with a parachute. A few days later *Geschwaderkommodore* Mölders fixed the *Eisernes Kreuz II. Klasse* on his chest.

On October 1, 1939, he was given the task of the deployment of the III. *Gruppe* of JG 53 in Wiesbaden-Erbenheim. His vigor and ability to lead were displayed by the fact that the new *Jagdgruppe* could report that they were ready for duty after 14 days, and were included in the border surveillance. On October 30, 1939, the *Gruppe* achieved their first air victory, and no one other than Werner Mölders was the lucky victor. It was an English Bristol-Blenheim that flew reconnaissance in the Trier region. When the German fighter pilots were finally allowed to fly over the border, opportunities for shootdowns increased. Until the beginning of the military campaign against France there were 18 air battles written down in his flight book, nine of which he could enter in as shootdowns—an astonishing ratio. Under the opponents that were shot down, four English Hurricanes of the 1st and 73rd Squadron were included. On April 2, 1940, he was awarded with the *Eisernes Kreuz I. Klasse* after seven air victories.

On May 10, 1940, the military campaign against France began. The first days of the offensive brought Mölders' *Jagdgruppe* hardly any contact with the enemy. On 14 May Mölders had the first air battle, and sent his 10th opponent, a Hurricane, to the ground. Successes came in quick succession. On 21 May the *Jagdgruppe* engaged in air combat with more than 50 English and French fighter pilots, and Mölders defeated three Moranes. Another day he, with his *Gruppenschwarm*, hunted down a French reconnaissance, a Potez 63, that skillfully defended itself. Finally, right above the treetops, Mölders was in ramming distance, and a fire on impact marked the end of this brave opponent. During the return it was learned that the tips of the Bf 109's propeller were bent, as he must have touched the treetops during the heated hunt. On the next day he carried out a successful low-level attack with ten Bf 109s in bad weather on the French Mourmelon air field. And on 27 May his 19th and 20th air victories were due. Mölders spotted six of the modern French Block-*Jäger* that unsuspectingly flew in the direction of the front. Skillfully, he went around the opponent in a wide circle and attacked him with his *Staffel* coming from the French hinterland:

"Move up, - Attack! Like a thunderstorm we came from above to the surprised opponent. I take aim at the right of the back *Kette*, *Leutnant* Panthen the left, *Leutnant* Müller pushes through towards the *Kettenführer*. - Bloch-*Jäger*! Very clearly I see the blue-white-red stripes on the rudder assembly - approaching meter by meter ... Meter, calmly aim and - the *Garbe*! My *Kanonen* and *MG-Garben* flash in the enemy crate. A French fighter flies apart again."

2 Curtisses according to the fight book, 2 Blochs according to personal description from Mölders in "Forell":

"My Mes are shooting out in all directions. The second *Schwarm* broke through the front *Kette*. Everywhere the 109 is sitting behind the Bloch. The enemy fighter pilots have not escaped any shot. - A fighter shot down by *Leutnant* Kunert crashes, burning, to the ground. Below I count four impacts already. The enemy *Staffelführer* circles desperately. I don't let him go anymore. On fire, he crashes below not far from the plane Kunert shot down. The last one was so shot up by *Leutnant* Panthen that he had to go in for a belly landing."

On May 29, 1940, *Hauptmann* Mölders' name was mentioned for the first time in the report from the *Oberkommando* of the *Wehrmacht*. As the first German fighter pilot, he was awarded with the *Ritterkreuz zum Eisernen Kreuz*, presented by Göring. The joy for this high award was great, especially with his men. Even though there was no time for a big celebration because the opponents in the air were not yet brought down, "*Vati* Mölders" was enthusiastically received during his return from Göring's headquarters, and the event was appropriately celebrated that evening! On the morning of June 5, 1940, the 24th and 25th opponents fell—but during the afternoon his series of victories was interrupted. During his 32nd air battle since the outbreak of war he was shot down:

"17:15 takeoff, again with a *Staffel*. We fly to Amiens, and time is almost up. There are planes above us, but we can't make them out. We climb to 7,000, - Mes! Somewhat lower and gradually towards home. - Then suddenly: six Moranes! I prepare for attack, in the middle of the attack I recognize two foreign Me-*Staffeln* that attack the same opponent from behind and above. They are closer to him, so I retreat and watch from above. They shoot much too early, there is the usual winding up, where a few Moranes valiantly prepare for combat. A Me crashes down, on fire, the pilot hangs onto his parachute.

I watch the battle for a while and then attack a Morane that is being fired at again and again by three Messerschmitts. Soon I have them in my sight - they swim away quickly but haven't had enough. Suddenly they pull up under me, I lose them under the surface. - There it is again to the side under me - Damn it! - This Morane is also still shooting - Certainly very far. I turn away to pull into the sun. They must have lost me, because they are changing course in the opposite direction and disappear to the south. Below two Mes are still grappling with a last Morane. I observe the battle that passes in a low-level flight, where the Morane escapes shots by constantly circling. A glance above to the back - full of circling Mes. I'm approximately eight hundred meters high - suddenly there is a loud bang and sparks throughout my cabin, and everything goes black. The throttle control is shot, the control lever is hit forward, it is going down vertically. - Out now - or else it's over... I grasp the ejection lever, the canopy flies off, - then my brave bird rears up and gives me the last chance to undo the seatbelt and to get out of my seat.

Free! Pull on the ripcord - I suddenly tear it off in my hand, - I'm suddenly hit by shock. I reach above - but - the parachute already opened.

And then it becomes very calm. Once more I see my plane, crashing out of control, the left surface torn up; just above the ground it rears up once more as if it does not want to believe that it could be conquered after 25 victories, and then vertically hits and completely burns.

I am hanging onto the parachute; I look for my opponent, but only Mes circle around me - only Mes! Very slowly I glide to the ground, and this ground is still occupied by the enemy - 60 km behind the front, west of Compiégne. I draw my pistol, cock it, and then put it in the pocket of my pants. Below me two farmers get their horses together and flee. After a short orientation of the terrain I recognize a small forested area, otherwise just fields. The ground suddenly comes closer, I squat - the impact relatively soft, I am at once free from the parachute and run towards the forest. The French come running along from the sides - already the edge of the forest - a shot goes by my ear. I throw my fur coat and run so that I become short of breath, to the end of the little forest...."

This air battle was viewed like this in the eyes of the opponent:

> "At 17:05 8 Dewoitine 520 fighter planes of the *Staffel* 'France' of the *Jagdgruppe* II/7, accompanied by a triple *Patrouille* of the *Jagdgruppe* I/3, that operates from a low altitude, a parachute mission in the Braye-sur-Somme, Proyart, and Athies region. Above Compiégne they were surprised at an altitude of 8,200 meters by 15 Me 109s and another unit of 25 planes. During the first attack by German fighter pilots the Dewoitine 520 of the *Oberstabsfeldwebel* Ponteins of the *Höhenpatrouille* on fire crashes. The injured pilot saves himself with a parachute. After an impetuous maneuver his *Rottenflieger* must stop the battle. Now the *Zwischenpatrouille* is being attacked. The Polish pilot *Leutnant* Strzembosh is badly injured, and *Leutnant* Louis crashes fatally with his Dewoitine 520 by Quesnel-Aubry. Under the leadership of *Hauptmann* Huo, the *Tiefpatrouille* achieves a bitter resistance. Two bursts of fire from *Leutnant* Pomier-Layragues set a Me 109 on fire, whose pilot gets out immediately. Taken prisoner by *Artilleristen* of the 3. *Gruppe* of the 195th *Fußartillerieregiment*, this pilot is none other than *Hauptmann* Werner Mölders. In the course of air combat with four opponents, *Leutnant* Pomier-Layragues brings down another Me 109 in Cauly. He is then hunted by six others that hit him hard. Like a torch the Dewoitine 520 crashes in the suburbs of Marissel with fire upon impact, where the pilot found his death."

A French description of his captivity exists that appears very typical:

> "An infantry unit, unloading on the Bahnhof Estrees St. Denis, saw the parachute floating down to the ground not far from Hof Villerseau. The pilot attempted to disappear into the surrounding cornfields. When he saw soldiers approaching him from all sides he let himself be taken captive. He was a lanky, lean, tanned guy, who seemed to maintain his composure even in the last minutes of the hasty events. Someone brought him to Schloß Blincourt. There Commandant Bassous and Captain Drouot questioned him with the help of an interpreter, *Korporal* Zimmermann. He was, as one clearly saw, of Prussian descent, and had nothing with him except a photograph that showed him by Göring's side. Cdt. Bassous gave him back the *Eisernes Kreuz* that was taken from him and squeezed his hand. Mölders asked to meet the French pilot who shot him down; he wanted to congratulate him. But Bassous responded that it was not possible, because he did not know which unit the plane belonged to."

This report conceals the fact that Mölders was hit with rifle butts and kicked, left bleeding, during captivity. He was lucky, because after the end of the military campaign against France it was learned that angry Frenchmen killed many German pilots. In a ten day Odyssey, Mölders was brought to the *Offizierslager* Monferran by Toulouse. There the treatment was excellent, if there was no concern that the French would give him over to the Englanders. After the armistice he returned home and was received by Göring.

It should be mentioned here that the Frenchman who mishandled Mölders the worst in captivity was sentenced to death after the surrender of France, but was then, after repeated pleas from Mölders, pardoned by Göring with the words: "Mölders, I give you the man!" However, Mölders looked at his experience as a one-time derailment, and spoke of his opponents always with great respect.

A three-week vacation at home followed; on 19 July he was promoted to *Major*, and on July 20, 1940, he was appointed to *Kommodore* of *Jagdgeschwader* 51. His predecessor was *Generalmajor* Theo Ostercamp, who took over the post as *Jagdfliegerführer* (Jafü). The *Geschwader*'s combat post was located in Wissant, in the Hotel Bellevue. The JG 51 carried the burden of air warfare against England on the channel at this time.

On July 28, 1940, Mölders flew his first mission as *Kommodore*. It did not proceed as practically as he had hoped:

"I flew with my *Adjutant*, *Oberleutnant* Kircheifl, north of Dover. Suddenly, I see three English fighter planes, and behind them a swarm of Spitfires in the smoke. The Spitfires fly a bit lower. I take on the *Kette*. As I near, both outermost Spitfires turn, the middle one flies on. I set myself behind him and shoot from a distance of 60 m. The right side catches fire immediately, thick smoke and flames come from the plane that disappears below. I pull up and see a swarm of Spitfires, 8-10, behind me. I am in great shock. Only one thing can help here: push through the middle of the swarm! I sweep through, the front planes cannot reach me any longer, but one behind me is watching carefully.

He shoots and hits. It is rattling inside my plane. Shots in the cooling system, sides, and gas tank. I retreat while the sides are turning and continue at 700 km per hour towards the channel. The Spitfires are after me and my trail of smoke. Thank God the motor holds out. Then *Oberleutnant* Leppla comes to help. He observed the fuss. He attacked the Spitfire that was ready to attack. After a few seconds he crashed below, covered in smoke. When I reached the coast the engine began to grumble. During landing the landing gear did not come out. I made a smooth belly landing. When I wanted to climb out of the plane, my legs were unusually weak. I saw traces of blood. The results of the examination in the military hospital: three fractures in the thigh, one in the knee joint, and one in the left foot. In the heat of combat I didn't notice a thing."

According to English documents, the lucky English scorer could have been Flight Lieutenant Webster from the 41st Squ., who did not survive the air combat over England and fell on September 5, 1940. The Spitfire shot down in flames by Mölders made a belly landing, and its pilot, Flying Officer Lovell, came into the military hospital wounded. Mölders also had to remain in the military hospital for 11 days. In his absence Osterkamp led the *Geschwader*. On 7 August Mölders returned, and just at the right time, because air combat came closer to its peak.

On August 26, 1940, Mölders began his new series of victories with the shootdown of a Spitfire by Folkestone. Just at the right time, because a number of fighter pilots, like Wick, Balthasar, Galland, Oesau, Mayer, and Tietzen almost caught up with him in their shootdowns. That should not mean that Mölders saw his mission as a fighter pilot as a kind of competition. The fulfillment of his duties were the guiding principle of his behavior; the war was a very serious thing for him, not to be devalued as a private affair, or on one's own account. If he came across such an attitude in his *Gruppe*, or in his *Geschwader*, he stepped in. If the public had the impression that there was a competition between Mölders, Galland, and Wick, or between single units, it was the primary responsibility of the war propaganda. Mölders was above such things.

On 31 August he defeated three Hurricanes, and another mention in the *OKW-Bericht* followed. And on 10 September he reached his 40th air victory with a shootdown of two Spitfires of the 92nd Squadron.

His *Rottenflieger* and friend, *Oberleutnant* Claus, who he recently received from his old *Jagdgruppe* III./JG 53, gave the following report:

"I saw a Spitfire *Staffel* above Dungeness and reported it. I received the answer: '*Eigene*.' When the Spitfire pulled through and I could clearly recognize the *Kokarde*, I had the courage and announced: '5 Spitfires behind us!' We made an about turn and set behind them. The Englanders turned to the left. Mölders set after a Spitfire that unsuspectingly flew straight ahead as a right '*Holzauge*.' Then he shot from a close distance. Then immediately burning below. Mölders switched to the next plane, the left man of the *Kette*. He looked to his *Staffelführer* and made a left turn. Mölders dropped behind him. The tail unit was smashed under his *Garben*, parts of the plane flew through the area. The other three circled wildly and disappeared far underneath the clouds. In the brook I saw a spurting fountain. The second Spitfire tumbled like a withered leaf below and hit the channel. There was no parachute jump from either plane."

Mölders was the first fighter pilot in World War II who reached 40 shootdowns. The next day he was awarded as the second *Offizier* of the *Wehrmacht* the *Eichenlaub zum Ritterkreuz des Eisernen Kreuzes*, that until now only *General* Dietl had received, and it was presented by Hitler in the *Reichskanzlei* in Berlin.

On October 1, 1940, his brother, *Oberleutnant* Victor Mölders, whom he received as *Staffelkapitän* in the *Geschwader*, was shot down in a *Jagdbomber* mission over England and was taken captive. Mölders single-mindedly led the way in the more and more difficult battles on the channel that involved heavy loses. Three Hurricanes fell on 12 October, and with three more Hurricanes on 22 October he won his 49-51st air victories, again the first. On October 29, 1940, his 54th opponent fell. Then a bad flu kept him on the ground, and for many weeks he could not fly. During this time his earlier *Rottenflieger*, who accompanied him on over 60 air battles, *Oberleutnant* Georg Claus, fell above the mouth of the Thames. Mölders desperately searched with his Bf 109, and circled over the water surface in vain until his last drop of gas. On November 28, 1940, Major Helmut Wick, *Kommodore* of the JG 2 "Richthofen," had to jump with his parachute above the channel, and remained missing after 56 air victories. Wick, who was one of his students with the 1. *Staffel* JG 53, once wrote about his time with Mölders:

"When I came to the 1./JG 53 as a young *Leutnant*, there I was, as one would say, the last speck of dirt. *Major* Mölders, bearer of the *Goldenes Spanienkreuz mit Brillanten*, then already the *As* of the *Asse*, made me what I am today. Mölders untiringly trained us, practiced with us, supported us by describing his experiences, air battles, and how they are in reality. When I was drilled after four months of '*Katschmarek* time' on the airfield, he gave me a few chances—he let me attack the enemy. I utilized these chances, and I am always thankful to Mölders for it! Mölders was my teacher who, for me, was always the epitome of a first-rate teacher, the best superior, and sincere comrade. He will always be a model for each fighter pilot that once flew under him. He is, at any rate, my model in each second of each mission. I would like to go as far to say: He is the person who trained me to be a useful *Offizier* and fighter pilot. The time in the *Gruppe* of Major von Janson and the *Staffel* of Hauptmann Mölders was crucial for my later successes."

After bad weather had forced a temporary cessation of warfare against England, the JG 51 was pulled out of action. Mölders (*Oberstleutnant* since October 25, 1940) and his pilots spent a few carefree weeks on a ski vacation in Vorarlberg. At the beginning of February 1941 the *Geschwader* went back to the channel. Already on February 10, 1941, Mölders proved that he did not forget anything, as a Hurricane crashed under his bursts of fire by Calais into the ocean. On 26 February his 60th opponent fell. When the *Geschwader* was pulled from the channel in June 1941 he was responsible for 68 air victories. His flight book listed 238 missions, to which 71 front surveillance flights, alarm starts, and advances were added. He was in air combat seventy times—68 air victories won—and nothing more clearly shows what kind of exception this man must have been among the fighter pilots.

Luck alone, as well as experience, are not explanations. A good eye, aviation capability, reactions, tactical talent, the art of shooting—it all must add up to reach such achievements. Apart from the fact that it was an achievement to have survived 70 air battles. How many fighter pilots were there who had just one, or maybe no shots after 70 air battles, and were good soldiers who committed themselves completely, as a *Rottenflieger*, for example, on whom one could depend in each critical situation.

In the course of June 1941 almost the entire German *Luftwaffe* transferred to the western Russian border. On the evening before the attack on Russia, Mölders spoke in Siedlce to the pilots of his *Geschwader*. He thought that the war against Russia would be very difficult and would last very long. First it did not look the sort. The Russian *Luftwaffe* suffered terrible losses in the first weeks of the military campaign, and no one thought it was possible that they could recuperate from it. On the first day of the offensive Mölders achieved four air victories, and on the same day he was awarded as the second soldier of the *Wehrmacht* with the *Schwerter zum Eichenlaub*, after Galland received this award a day earlier. Things started happening very fast. On June 30, 1941, his *Jagdgeschwader* shot down 110 Russian planes, five of which fell to Mölders. On 3 July 3 he remained in the *Führer*'s headquarters "Wolfsschanze," where Hitler presented him the *Schwerter zum Eichenlaub*. Two days later Mölders was on a mission again, and extended his shootdown list with two Russian bombers and two fighters. He had long since exceeded the magic number 80, which marked the shootdown record of the legendary Manfred von Richthofen from World War I. And on July 15, 1941, Werner Mölders set a new magic number when he achieved his 100th and 101st air victories, and was spinning with delight.

The next day it was announced that the highest German award for bravery, the *Eichenlaub mit Schwertern und Brillanten zum Ritterkreuz des Eisernen Kreuzes*, was awarded to him as the first soldier on July 15, 1941. On 20 July he was promoted to *Oberst*—at only 28 years old! His joy was spoiled by the fact that he received a grounding order, and had to give up the command of his *Jagdgeschwader* to transfer to the *Reichsluftfahrtministerium*. During his departure he thanked his men for their loyal following, and wished everyone further luck in war. "I'm going into the dreadful hunting grounds of red tape," he said. On 26 July Hitler received him in the "Wolfsschanze" and presented him the *Brillant*. A meteoric career reached its height. He had reached his 101st air victory in 291 missions. Along with these were the missions in Spain and his 14 confirmed shootdowns.

On August 7, 1941, he was appointed to *Inspekteur* of the fighter pilots. Though his departure from the Front was difficult, his new position corresponded to his inclinations and ability. He once said to Galland:

> "You can be the Richthofen of the *Luftwaffe*. But I would prefer to be your Boelcke one day."

With the great fighter pilot of World War I in mind, he hoped to have a significant impact on his *Waffe* in the new sphere of activity, in which he would exercise his influence on the organization, the equipment of the units, technical developments, the development of tactics, and the bettering of training and leadership. This was an enormous program that required all of his energy to carry out.

In his new position Mölders had insight that he did not have earlier. Soon he recognized the shortcomings of the *Luftwaffe* in production, in supply of fuel and munitions, and in plane development. He also recognized that the battle against the Soviet Union would last a long time, that Hitler repeatedly took action against the advice of military experts, and that Göring, as *Oberbefehlshaber* of the *Luftwaffe*, had taken on too much. Mölders began to be greatly concerned about the future of the German people. It seems permitted to say that his early death saved him difficult moral conflicts that he inevitably would have dealt with.

But first he celebrated his marriage with *Frau* Luise Baldauf, the widow of a friend who died in a fatal crash as *Fliegerhauptmann*, on 13 September on Burg Falkenstein, in Taunus. His *Adjutant*, Hartmann Grasser, and his *Rottenflieger*, Erwin Fleig, were the witnesses to the marriage. The next weeks passed with inspection trips to fighter units. Then he took over the tactical leadership of the *Jagd* and *Schlachtflieger* on the Krim Peninsula as so-called *Nahkampfführer*. It was no wonder that the immediate proximity to the front did not let him rest. To the concern of his *Stab* he prepared a Bf 109 whenever he had the opportunity and flew missions, mostly in the accompaniment of *Unteroffizier* Günther Behling from the III./JG 77. Missions over Sewastopol and the Kertsch peninsula were confirmed for 8 and 13 November 1941. He achieved at least three air victories during these excursions that, naturally, could not be acknowledged.

On November 17, 1941, the *Generalluftzeugmeister*, *Generaloberst* Ernst Udet, committed suicide. *Oberst* Mölders received the command to join the funeral ceremony in Berlin. Bad weather prevented a timely departure. Only on 21 November could Mölders and his *Adjutant*, Major Wenzel, take-off as passengers of a He 111. This was navigated by *Oberleutnant* Kolbe, an experienced *Kampfflieger* of the KG 27 "Boelcke." The flight went over Nikolajew to Lemberg, where they spent the night. The next morning they continued. Between Lodz (then Litzmannstadt) and Breslau the left engine stalled. *Oberleutnant* Kolbe decided to fly through to Breslau instead of landing on the next airfield. It must remain a supposition that Mölders influenced this decision. When the plane got closer to the Breslau-Gandau airport, the radio operator, *Oberfeldwebel* Tenz, announced: "Single-engine flight. *Oberst* Mölders on board." The people gathered to receive them and heard the plane fly over the area two times but, despite the beaming sun, did not see the plane due to heavy ground fog. Even though Gandau and the nearby field, Schöngarten, had turned on the beacons, and continually fired "green" as a signal for landing clearance, the He 111 had ground contact just during the third approach. In the meantime it dropped a great deal in altitude, hung over the sick motor, and became tail-heavy. It was too late for a parachute jump. Kolbe pulled the plane over a cable railway that covered a brickyard. As the grounds of the brickyard emerged from the fog, he tried to pull up once more or to turn away, but in this phase the right motor stalled. The He 111 crashed and hit the embankment with its left wing. It turned 180 degrees and broke into two pieces.

As helpers broke open the cockpit, they found the lifeless body of a highly decorated *Offizier*. Someone raised him from his seat, but he was already dead. *Oberst* Mölders died from a fracture of the spinal column and contusions on his rib cage. Had he survived if he had worn a seatbelt is a useless speculation. The pilot, *Oberleutnant* Kolbe, was buckled in, but he passed away during the ride to the military hospital, as did the flight mechanic, *Oberfeldwebel* Hobbie. The radio operator suffered only an ankle fracture. *Major* Wenzel survived, as well, but with arm and leg fractures and a severe concussion. On November 25, 1941, the coffin with Werner Mölders' mortal remains was led through Breslau in an impressive funeral procession with many participants. The procession also went by Manfred von Richthofen's birthplace. Then the transport to Berlin followed. Hitler arranged a state funeral. The *Jagdgeschwader* 51 received the name "*Jagdgeschwader* Mölders" in recognition of *Oberst* Mölders' services, and its members could wear the armband with the same name.

On November 28, 1941, the ceremonious act of state took place in the ceremony hall of the *Reichsluftfahrtministerium* in Berlin at 11:00. Subsequently, the coffin was transported in a funeral procession through the city to the *Invalidenfriedhof*. There Werner Mölders was buried next to Ernst Udet and Manfred von Richthofen, both the most successful fighter pilots of World War I. In 1975 the government of the GDR had the graves of Mölders, Udet, and others in the *Invalidenfriedhof* leveled out.

Those born after the war will find it hard to understand the emotions of the majority of the Germans on the occasion of the death of Werner Mölders. Werner Mölders was undoubtedly one of the most popular figures during the years of the war, comparable with maybe Erwin Rommel and the *Unterseebootkommandant* Günther Prien. In these times of war he was the embodiment of the brave, chivalrous, proper soldier and *Offizier*, exaggerated in fantasy as a daring air hero, of a "knight in the skies." As the news of his death became known, many people felt as if paralyzed. November 22, 1941, caused people to think that the time of the special report-fanfare had come to an end. The death of Werner Mölders caused genuine sadness. The news of the puzzling death of the popular Ernst Udet had just come over the air—and now Werner Mölders....

Nothing reflects the fascination of his person better than the fact that shortly after his death, peculiar rumors were making their rounds, from disappearing into a concentration camp to a Catholic martyr. Today we know to what extent these rumors were being steered from the opposite side. And they hit the ground.

In the *Vierteljahresheft für Zeitgeschichte, Heft 1*, January 1968, Helmut Witetschek published an investigation of the rumors about Werner Mölders, from which we quote extracts:

"Since January 1942 a letter was spread around that *Oberst* Werner Mölders supposedly wrote to the Catholic priest Johst von Stettin, shortly before his death. The letter contains his clear profession to the Catholic church. While Mölders, who was part of the Catholic youth movement, was known as a practicing Catholic, much of the population did not doubt the authenticity of the letter. Above all, the religious groups of both confessions duplicated the letter and circulated it. Many clergymen of both Christian professions in all of Germany read the letter from the cockpit during church service. In April 1942 the *Staatspolizeistelle* Nürnberg-Fürth determined, for example, that alone in their area eleven catholic and seven evangelist clergymen and nineteen catholic and seven evangelist laymen had spread the letter, or read it during church service.

The *Gestapo*, which investigated the authenticity of the letter, made Werner Mölders' mother make a statement that the letter was not written in her son's style and, therefore, could not have been written by him. The incumbent Probst von Stettin, Ernst Daniel, read an announcement on Sunday, February 8, 1942, in the Propsteikirche St. Johannes Baptist in Stettin, in which he clearly stated that there never was a Propst Johst in Stettin, that he never personally knew Mölders, never received a letter from him, and never wrote one to him. The Catholic Feldbischof, Franz Justus Rarkowski, included an announcement in the *Verordnungsblatt* of the Catholic *Feldbischof* in which the armed forces clergymen pointed out that the letter was a crude counterfeit. The *Gestapo* took action against the spreading of the letter.... But the speed with which the police and administrative authorities depicted the letter as a fake made a large percentage of the population skeptical, especially while the letter seemed to confirm numerous rumors.

Afterwards Mölders did not carry out any missions in protest against the anticlerical measures of the regime. He was not killed by an accident, but by the SS, because he was in the way of the NS-*Führung* as a Catholic; he took the *Bischof* von Münster from the concentration camp; Hitler did not observe him during a reception because of his confessional stance, so he committed suicide out of melancholy....

The question of the letter's author could be clarified by the autobiography from Sefton Delmar, who identified the Mölders letter as a forging of the English secret service...."

The fact is that Mölders was good friends with the director of the Catholic *Jugendseelsorgeamt* in Berlin, *Pfarrrer* Erich Klawitter, who was, as a young *Kaplan*, Mölders' religion teacher and his school friend, the later *Erzpreister* Heribert Rosal.

Both school friends were members of the Catholic *Neudeutschland-Gruppe* in Brandenburg that Mölders led for a long time. Erich Klawitter carried out Mölders' religious wedding. Klawitter read a letter at the catholic youth day in Stettin that Mölders wrote to him on October 6, 1940. Without a doubt he wanted to show his young listeners how valuable it was to have a Christian foundation during these difficult times. This letter from Werner Mölders contained the following wording:

"On the Field, October 6, 1940

Dear Herr Direktor!

I was very happy to receive your good wishes. I have thought about the times I spent with you in which you brought, with great understanding, the true, natural Christian faith into my young soul. I have often thanked God for it, and know that I act in his divine providence. I have constantly been taught your work by Heribert Rosal, and hope to make my wish of visiting you come true. I could then tell you much of my life. A quick, successful life of a soldier, which has given me the opportunity to see the world during the military campaign in Spain, and the stay in Rome and Athens. I have experienced much joy, and I can tell you today that I am as you knew me. I will continue to do my duty, and thank you from the bottom of my heart for your prayers for the *hl. Opfer*, in which I put my trust completely.

Thankfully yours,
Werner Mölders"

Mölders' Catholic beliefs were known to the English secret service through prisoner statements, and the many rumors of his death were also known. The opportunity to stir up excitement in the German population was taken advantage of. On the broadcasting station "Gustav Siegfried Eins" of the English secret service, Heinrich Himmler was accused of having cowardly murdered Mölders, the "shining model of German masculinity." Sefton Delmer, one of the leading English secret service men, forged a letter that was supposedly from Mölders. This letter contained, from a clear profession to Catholic beliefs, a few remarks on death and the sense of war that one could classify as defeatist. The forged letter was shaken off as a leaflet in Germany.

A few words on the rumors on connections between Werner Mölders and the Bischof von Münster, Clemens August Graf von Galan. After the situation with the document there were no clues until now that Mölders personally intervened with any kind of military or political personalities in order to protect from possible political persecution. It is correct that Mölders was concerned with Galen's sermons, the anticlerical measures, and on euthanasia, expressed to the Bischof by a mediator. It coincides with the account from *Oberst a.D.* Johannes Janke, who knew Mölders very well. Afterwards, Mölders was not in agreement with National Socialism concerning the questions on the persecution of the Jews, nor the anticlerical proceedings. That was a result of his views and upbringing. He was an *Offizier* of old Prussian character, who obtained his views from his parents' house and from the *Gymnasium*. For him, service was of utmost importance; therefore, he became a soldier. And this let him endure many things and internally process what did not fit into his philosophy of life. It is moving to know how much admiration everyone that knew him shows for him after many years. His cousin Fritz von Forell wrote:

"In this man one senses his outstanding traits: the seriousness of his philosophy of life; his comradeship and goodness; his readiness to help, and restless devotion to something that he values as right and just; his serious, but often happy nature; his dry humor, that was with him in the most difficult of situations; his loyal faith in God, to whom he was responsible to; and finally, his absolute strictness with himself. He was one rare man, who understood the word duty as a natural thing."

Günther Rübell, once a young *Offizier* in his *Geschwader* wrote:

"The very sensitive Werner Mölders has a fine sense a being a *Truppenführer*, to get the proper men, who corresponded to his nature, to leading positions in his *Geschwader*, and at the same time replace overaged, 'old-established' *Kommandeure* with capable, young *Offiziere*. With the appointed men he formed a *Geschwader*, to which he was and remained "*Vati* Mölders" for his aviation efforts and successes. Being helpful always came before his orders, commands, and also tactical directions for air battle, because Werner Mölders wanted to train a multitude of good fighter pilots on a broad basis, instead of establishing single '*Asse*.' Highly decorated *Offiziere*, whose selfish personal striving stood in contrast to the leadership and nature of Werner Mölders, were also replaced. Nevertheless, he formed a collective, because such contradicted the original establishment of single *Einzelkämpfer*. The intellectual fulfillment of duty in lively comradeship united these individualists to a community under his leadership, whose basic attitude: 'One for all - all for one,' was lasting."

Werner Mölders was, naturally, not immune to judgement and assessment of his mistakes—who is perfect? He later voiced his opinions and admitted to his mistakes. His striving for the truth—also in himself—found expression herein: he stood against objective criticism, was always open-minded, and it prompted him to further constructive measures that were planned in his last memorandum, pointing the way towards the future. He was suited for it because he rejected opportunism that was eager for applause. It was this trait that gave him his great self-assertion of his ideas in his leadership of the *Luftwaffe*. People listened to him. So his early death did not only affect the *Geschwader* and the *Jagdfliegerei*, but also the entire *Luftwaffe*. Günter Rübell quotes statements from Hartmann Grasser, Mölders' *Adjutant*:

> "Hartmann later admiringly spoke much of his mentor's circumspection of the tactical flight plans. Again and again he was amazed by the early detection of planes in the air by Werner Mölders, who flew the best line of departure from a high elevation with the sun on his back in *Jagdkamp*—without being discovered or seen himself—to reach the opponent with a surprise attack and shoot him down with aimed, short bursts of fire. After each mission a portrayal of the route with plane movements followed in order to detect errors that could be prevented as a result of discussion afterwards. Hartmann Grasser was with Mölders for months on the ground, as well as in the air, and through many personal discussions—beyond wartime missions—became trusted by the *Kommodore* who, in a father-like manner, called him 'little one' and expressed their friendly relationship."

There is more to be said about Werner Mölders, the man. His reaction was typical, as he observed how one of his *Offiziere* fired at a railway train with aircraft weapons during the air battle above England. Mölders ordered the *Offizier* to come to him, and with a hot temper explained to him the difference between military and civilian goals. And Johannes Janke explained:

> "Mölders always had an open ear for the worries and needs of his inferiors. One could at any time go to him with their worries. - Even when higher superiors were with him. 'Herr Major? I have a question!' He promised not only to help, but then he actually did."

Considering all these statements, it goes without saying that the *Bundeswehr* found Werner Mölders on a search for role models from German military history. Whoever reads his thoughts on the topic "Command and Obedience," which could be described as revolutionary from the point of view back then, will understand this selection:

"...Authority may not be based on power. It is a concern of the moral and spiritual superiority of the superiors. The pride of one's own flawlessness must outshine the feeling of the necessary subordination in the soldier...

It is easy to say: You must blindly obey each order and not question anything. Behind the command is the power of your superiors that will break disobedience. The militarists embody this view.

On the other hand, a good superior will instruct his soldiers: Without obedience a troop cannot last. Your superiors can force you to be obedient, but I would like for you to voluntarily submit to the conviction that it must be so. Obedience is voluntary discipline. Out of respect for your uniform and out of respect for yourselves you should obey. The military march its negative spin-offs of exaggerated tautness is tedious army life and reprehensible. We need self-chosen order. We want to see flawlessly trained soldiers on the street, but no wiry figures clicking their heels together. The salutation is natural, as well as among civilians. But the superiors should really be worthy of the salute. And without *Tjawoll* and twisting of the arm. It is no threat to discipline to go out in civilian clothes. Why should the soldier not have the pleasure of being able to go out without danger, with his hand at his cap though the streets? Why should a theft or other crime committed by a soldier be worse than that of a civilian? And does the soldier really need a special right to complain, and must the form of address be in third person? To secure the proper respect for the superiors? In the civilian life this form of address is not known, but whoever deserves this respect, deserves it!

Man should always first remain a man in uniform! The drilling of the troops is militarily necessary. But it will be made worse if it becomes an end in itself.

When an idiot of an instructor drives me up the wall and shouts: 'I am the dumbest so-and-so of the entire army,' then that is diabolical sadism, and has nothing to do with being a soldier—it is militarism. I never tolerated such negative spin-offs. And are my people bad soldiers?

Order and punishment must exist. We do not get around it in any community. But order must be based on voluntary discipline. And with reason. It is acquired. Discipline is voluntary subordination from the realization of the essential...."

Mölders stops here with the words of the great philosopher Kant, who said:

"The ability to create the motivation for wanting, that is freedom."

On November 9, 1972, the name "Mölders" was given to the *Kaserne* of the II. *Abteilung* of the *Fernmelde-Regiment* 34 of the *Bundesluftwaffe* in Visselhövede.

On the 32nd anniversary of Werner Mölders' death *Jagdgeschwader* 74 of the *Bundesluftwaffe* in Neuburg/Donau was given the name "Mölders." Mölders is represented in two sections of the *Bundeswehr*, Marine and *Luftwaffe*. *The Bundeswehr must be proud to keep the memory of this man alive*, who remained true to the Prussian-German tradition of fulfilling his soldierly duty, but who attempted to loosen his rigid behavior and decisions as *Truppenführer*, and proved his courage before the thrones of princes.

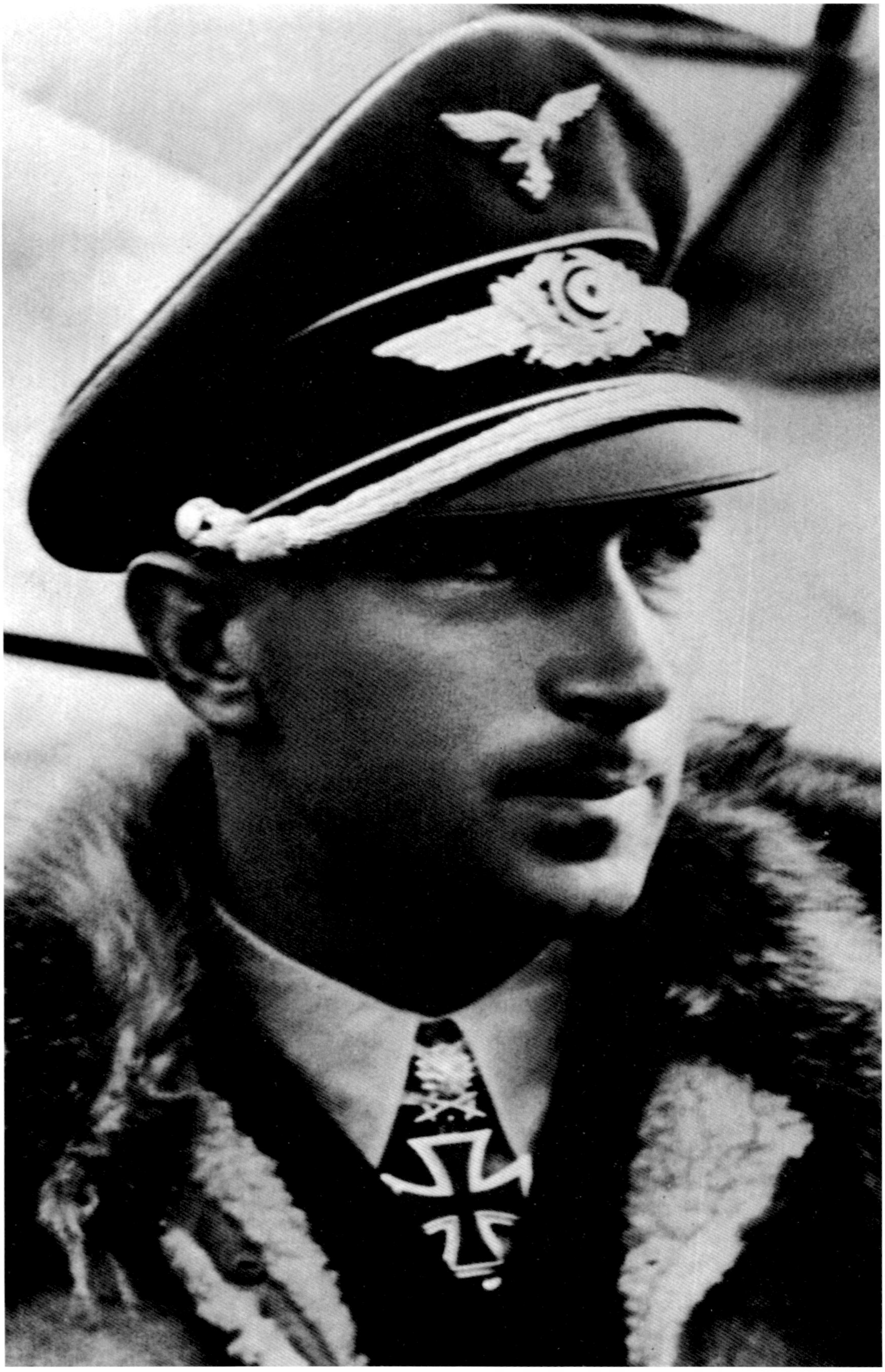

A Timeline of Werner Mölders' Life

Born on March 18, 1913, in Gelsenkirchen/Westfalen as the son of the secondary schoolteacher, Viktor Mölders, and his wife Annemarie Mölders, born Riedel. Secondary schoolteacher, Viktor Mölders, fell on March 2, 1915, as *Leutnant* of the Reserves in an *Inf. Rgt.* in L'Argonne (France).

1919-1931
Attendance of the *Volksschule* and the "Saldria-Gymnasium" in Brandenburg/Havel; completion of his school-leaving exam. As a student he was an excellent athlete with a fondness of rowing, and a love for nature and hunting.

April 1, 1931
Entry as *Offiziersanwärter* in the *100,000-Mann-Heer* of the *Reichswehr*, II./*Inf. Rgt.* 2 in Allenstein/East Prussia.

October 1, 1931
Promotion to *Fahnenjunker-Gefreite*.

April 1, 1932
Promotion to *Fahnenjunker-Unteroffizier*.

October 1932
Attendance of the *Reichswehr*'s *Fähnrich* course at the *Kriegsschule* in Dresden.

June 1, 1933
End of the course and promotion to *Fähnrich*.

June 11, 1933
Transfer to 1. (Prussian) *Pionier-Bataillon* (*Inf. Regt.* 2) at the *Pionierschule* in München.

February 1, 1934
Promotion to *Oberfähnrich*.

February 6, 1934
Until December 31, 1934, pilot training at the *Deutsche Verkehrsfliegerschule* in Cottbus (Mark Brandenburg).

March 1, 1934
Promotion to *Leutnant*.

January 1, 1935
Until June 30, 1935, further pilot training at the *Kampffliegerschule* Tutow. Meanwhile, completion of his training as a fighter pilot at the *Jagdfliegerschule* Schleißheim.

March 1, 1935
Lifting of the *Tarnbestimmung* for the third *Wehrmachtsteil*, the *Luftwaffe*, and official takeover as *Leutnant* in the new German *Luftwaffe*.

May 21, 1935
Conferment of the *Flugzeugführerabzeichens* of the *Luftwaffe*.

July 1, 1935
Until March 31, 1936, with the "*Fliegergruppe* Schwerin" (I./162) "Immelmann."

March 7, 1936
Mission with his *Staffel* during the occupation of the demilitarized Rheinland. First unit to land in Düsseldorf.

April 1, 1936
Until March 14, 1937, head of the *Jagdschulstaffel* of the II. *Gruppe/Jagdgeschwader* 134 "Horst Wessel" in Werl/Westfalen.

April 20, 1936
Promoted to *Oberleutnant* as a result of April 1, 1936.

March 15, 1937
Until April 13, 1938, *Staffelkapitän* of the 1. *Staffel* in I./JG 334 in Wiesbaden (later renamed I./JG 133 and I./JG 53 "Pik As").

April 14, 1938
Until December 5, 1938, ordered to 3./J 88 of the "Legion Condor" during the civil war in Spain ("‹bung Rügen"). Mission on the Valencia-Ebro Front.

May 24, 1938
Taking-over of the *Staffel* (3./J 88) as *Staffelkapitän* for his predecessor, *Oberleutnant* Galland, who returned home.

July 15, 1938
Mölders' first air victory in Spain. Shootdown of a Curtiss in the Algar region.

October 18, 1938
Promoted to *Hauptmann* as a result of October 1, 1938, and his outstanding efforts as a superior and fighter pilot. His soldiers call him "*Vati* Mölders!"

November 3, 1938
Shootdown of a "Rata" in the Mola region. It was his 14th and last air victory in Spain.

December 5, 1938
Return to Germany. With 14 acknowledged air victories, Mölders is the most successful fighter pilot of the "Legion Condor."

German Fighter Ace Werner Mölders - An Illustrated Biography

December 6, 1938
Until March 1939 *Stabsarbeit* with the *Inspekteur* of the fighter pilots in the *Reichsluftfahrtministerium* in Berlin. (Working on new fighter pilot rules) Officially a member of I./JG 133 since December 6, 1938.

March 1939
Until October 1, 1949, *Staffelkapitän* of the 1. *Staffel* in I./JG 133 (later renamed I./JG 53 "Pik As").

May 4, 1939
Conferment of the Spanish awards "Medalla Militar" and "Medalla de la Campaña."

June 6, 1939
Conferment of the *Deutsches Spanien-Kreuz in Gold mit Schwertern und Brillanten*. On the same day took part in a huge reception for all highly decorated members of the "Legion Condor" in the new *Reichskanzlei* in Berlin.

September 20, 1939
Mölders' first air victory in World War II (incorrectly entered into his flight book under September 21, 1939). Shootdown of a French fighter plane, Curtiss, by Sierck. Conferment of the *Eisernes Kreuz 2. Klasse*.

October 1, 1939
Until June 5, 1940, *Kommandeur* of the III. *Gruppe* of the JG 53 "Pik As" (Deployment and taking-over of the *Gruppe* in Wiesbaden-Erbenheim).

April 2, 1940
7th air victory (a Hurricane). Conferment of the *Eisernes Kreuz 1. Klasse*.

May 10, 1940
Military campaign against France. First large-scale operations of the German *Luftwaffe* against western allied air forces.

May 27, 1940
19th and 20th air victories as the first German fighter pilot in World War II.

May 29, 1940
Conferment of the *Ritterkreuz zum Eisernen Kreuz*. Mölders was honored as the first German fighter pilot. First mention in the *Wehrmachtbericht* (*OKW-Bericht*).

June 6, 1940
In the Beauvais-Compiégne region Mölders achieved his 24th and 25th air victories in the morning. Around 18:40 of the same day he was shot down during his 133rd mission in the Compiégne region in air combat with French fighter planes (Dewoitine 520s) from *Leutnant* Pomier-Layragues, and was taken into French captivity after a parachute jump.

June 30, 1940
Return from captivity.

July 1, 1940
Until July 19, 1940, official member of the *Ergänzungs-Jagdgruppe* Merseburg.

July 19, 1940
Promoted to *Major*.

July 20, 1940
Until July 19, 1941, *Kommodore* of the JG 51. From July 20, 1940, until June 15, 1941, missions on the

channel front with his *Geschwader*, and from June 22, 1941, until July 19, 1941, missions in Russia.

July 28, 1940
First mission on the channel above England. After the shootdown of a Spitfire over Dover, pursued by six Spitfires, forced landing and, due to a bullet fracture on his leg, returned wounded.

August 1940
Conferment of the *Gemeinsames Flugzeugführer- und Beobachter-Abzeichen in Gold mit Brillanten* that is only awarded in few, special cases.

September 18, 1940
JG 51's 500th shootdown by Erwin Fleig.

September 20, 1940
By Dungeness shootdown of two Spitfires of the 92nd Sqn. and the 39th and 40th air victories. First fighter pilot of World War II who reached this many shootdowns.

September 21, 1940
Conferment of the *Eichenlaub zum Ritterkreuz des Eisernen Kreuzes* as second solder of the *Wehrmacht* (after General Dietl).

September 23, 1940
Presenting of the "*Eichenlaub*" by Hitler in the *Reichskanzlei* in Berlin.

October 22, 1940
Mölders achieved three shootdowns (No. 49-51) in air combat between six Bf 109s and 15 Hurricanes of the 74th and 605th Sqn. NW Maidstone. He was the first German fighter pilot to reach his 50th air victory.

October 25, 1940
On the occasion of his 50th air victory he was promoted to *Oberstleutnant* for his bravery and great service to the fighting strength of the German fighter pilots.

February 26, 1941
60th air victory reached. Shootdown of a Spitfire at 18:37 SO Dungeness at an altitude of 6500 m.

May 20, 1941
End of his service in the west on the channel. A total of 238 missions completed (309 including Front surveillance, reconnaissance, advances, etc.), 68 air victories achieved (16 French, 52 Englanders).

June 22, 1941
Four shootdowns on the first day of the military campaign against Russia. (69-72nd air victories). On the same day conferment of the *Eichenlaub mit Schwertern zum Ritterkreuz des Eisernen Kreuzes* as second solder of the Wehrmacht.

July 3, 1941
Presenting of the award by Hitler in the *Führer*'s headquarters, "Wolfschanze."

July 12, 1941
Since the beginning of action in the east, the JG 51 under Mölders' leadership shot down 500 Soviet planes with only three loses of their own until 12

July 12. On 12 July the *Geschwader* won their 1200th air victory.

July 15, 1941
Mölders reached his 100th and 101st air victories on this day! He was the first fighter pilot in the world to reach this number of shootdowns. Together with the 14 air victories in Spain he achieved a total of 115 air victories. As the first soldier of the *Wehrmacht*, Mölders was awarded with the *Eichenlaub mit Schwertern und Brillanten zum Ritterkreuz des Eisernen Kreuzes* (the highest German honor until December 29, 1944).

July 20, 1941
Promoted to *Oberst* (at only 28 years old!). At the same time a grounding order and transfer to *Reichsluftfahrtministerium* (RLM) in Berlin until August 6, 1941.

July 26, 1941
Hitler receives Mölders in his headquarters "Wolfschanze" (a few kilometers east of Rastenburg, in East Prussia), and presents him with the "*Brillant*."

August 7, 1941
Appointment to "*Inspekteur* of the fighter pilots." During inspection trips to the *Jagdgeschwader* he flew missions on the Eastern Front, despite the grounding order, and achieved an unknown number of shootdowns.

September 13, 1941
Wedding on Burg Falkenstein, in Taunus, with *Frau* Luise Baldauf, born Thurner.

November 17, 1941
Suicide of the *Generalluftzeugmeister*, *Generaloberst* Ernst Udet, in Berlin.

November 22, 1941
Fatal crash with a twin-engine courier plane Heinkel He 111 of the *Kampfgeschwader* 27 "Boelcke" by Breslau-Schöngarten on the flight from Cherson (Ukraine) to Berlin for Udet's funeral. Along with Mölders, *Oberleutnant* Kolbe, pilot, and *Oberfeldwebel* Hobbie, flight mechanic, also died. *Jagdgeschwader* 51 received his name in recognition of *Oberst* Mölders' services. At the same time his state funeral was being arranged.

November 28, 1941
At 11:00 am ceremonious act of state in the ceremony hall of the *Reichsluftfahrtministerium* in Berlin. Subsequent funeral procession through the city to the *Invalidenfriedhof*, where Mölders was buried next to Udet and Manfred Frhr. von Richthofen, the most successful fighter pilots of World War I.

April 13, 1968
Launching of the *Lenkwaffenzerstörer* D 186 of the *Bundesmarine* with the name "Mölders" in Bath/Maine (USA).

November 9, 1972
Conferment of the name "Mölders" to the *Kaserne* of the II. *Abteilung* of the *Fernmelde-Regiment* 34 of the *Bundesluftwaffe* in Visselhövede (Lüneburger Heide).

November 22, 1973
On the 32nd anniversary of Werner Mölders' death, conferment of the name "Mölders" to *Jagdgeschwader* 74 of the *Bundesluftwaffe* in Neuburg/Donau. Mölders is the only soldier in the history of the German army to have his name represented in two sections of the armed forces, the Marines and the *Luftwaffe* of the *Bundeswehr*.

1975
The GDR had all graves in the *Invalidenfriedhof*, the treasure of Prussian-German history in East Berlin, leveled out, as well as Mölders' grave.

The Mention of Werner Mölders and his *Geschwader* in the *Wehrmachtbericht* (*OKW-Bericht*), in Special Announcements and Press Reports.

May 29, 1940
OKW-Bericht: On 28 May the opponent's loses in the air carried a total of 24 planes, 16 of which were shot down by flak and eight in air combat. Three German planes are missing. *Hauptmann* Mölders reached his 20th air victory.

September 6, 1940
OKW-Bericht: Out of the four already named *Offiziere*, three other fighter pilots achieved 20 and more air victories in the last weeks: *Hauptmann* Mayer, *Hauptmann* Oesau, and *Hauptmann* Tietzen. At the top of all the victors in air combat is Major Mölders with 32 shootdowns.

September 23, 1940
OKW-Bericht: Major Mölders' *Jagdgeschwader* has, until now, achieved over 500 air victories.

September 25, 1940
OKW-Bericht: *Major* Mölders and *Major* Galland achieved their 40th air victory.

October 23, 1940
OKW-Bericht: *Major* Mölders shot down, as was already reported, his 49th, 50th, and 51st opponents in air combat against enemy fighter pilots that were superior in number.

October 26, 1940
OKW-Bericht: In the course of yesterday's air combat our fighter planes took out 17 enemy fighter pilots. With this *Oberstleutnant* Mölders achieved his 52nd and 53rd air victories. Nine of our own planes are missing.

February 11, 1941
OKW-Bericht: After yesterday and the following night the enemy carried a total loss of 33 planes. Two of our own planes are missing. *Oberstleutnant* Mölders achieved his 56th air victory.

February 27, 1941
OKW-Bericht: Between 23 and 26 February the German *Luftwaffe* destroyed 33 enemy planes, 18 of

which were in air battles, and three by anti-aircraft artillery, the rest were destroyed on the ground. During the same time ten of our own planes were lost. *Oberstleutnant* Mölders achieved his 60th air victory.

April 18, 1941
OKW-Bericht: *Oberstleutnant* Mölders achieved his 64th and 65th air victories on 16 April, *Oberstleutnant* Galland his 59th and 60th on 15 April.

June 23, 1941
OKW-Bericht: *Oberstleutnant* Mölders achieved his 72nd air victory yesterday.

June 24, 1941
OKW-Bericht: Under the leadership of *Oberstleutnant* Mölders, the *Jagdgeschwader* achieved its 750th air victory on June 22nd.

July 1, 1941
OKW-Bericht: In the victorious air battles in the east, the *Jagdgeschwader*, under the leadership of *Oberstleutnant* Mölders, achieved 110 shootdowns, the
Jagdgeschwader, under the leadership of *Major* Trautloft, 65 shootdowns.
Oberstleutnant Mölders achieved his 82nd, *Hauptmann* Joppien his 52nd air victory.

July 16, 1941
Special report: During the battles on the Eastern Front yesterday, *Oberstleutnant* Mölders, *Kommodore* of a *Jagdgeschwader*, shot down five Soviet planes. With that he achieved 101 shootdowns, including 14 shootdowns in the military campaign against Spain, and gained 115 air victories. The leader and *Oberster Befehlshaber* of the *Wehrmacht* awarded this heroic role model of the *Luftwaffe* and most successful fighter pilot of the world as the first *Offizier* of the German *Wehrmacht* the highest German medal of bravery, the *Eichenlaub mit Schwertern und Brillanten zum Ritterkreuz des Eisernen Kreuzes*.

July 16, 1941
OKW-Bericht: As already reported in the special report, *Oberstleutnant* Mölders, *Kommodore* of a *Jagdgeschwader*, shot down five more Soviet planes, and achieved with this his 101st air victory in this war.

July 17, 1941
Press report: Since the beginning of action until 12 July, Mölders' *Jagdgeschwader* shot down 500 Soviet planes with only three loses of their own. The *Geschwader* won their 1200th air victory on 12 July.

September 12, 1941
Press report: Mölders' *Jagdgeschwader* achieved its 2000th shootdown on 8 September. *Major* Beckh gained the 2001st air victory of the *Geschwader* through a shootdown of a Soviet fighter pilot. Until 10 September Mölders' *Jagdgeschwader* shot down a total of 2033 enemy planes, 1357 of which were in the east. 188 planes were destroyed by aircraft weapons from the ground, and 110 planes by bombs on the ground. 142 *Panzerkampfwagen*, 16 heavy artillery, 16 engines, 432 LKWs, 75 automobiles of all sorts, and a *Panzerzug* were destroyed. 16 bearers of the *Ritterkreuz* belong to this *Jagdgeschwader*!

November 22, 1941
Special report: A cruel twist of fate would have it, that several days after the death of the pilot hero from the World War, *Generaloberst* Udet, the German *Luftwaffe* lost the boldest and best of their young fighter pilots: *Inspekteur* of the fighter pilots, *Oberst* Werner Mölders, fatally crashed by Breslau on a service flight with a courier plane on 22 November. Undefeated by the enemy, the victor tragically found a tragic pilot's death after 115 air battles.

The achievements and successes of this glowing fighting spirit, just 28 years old *Offizier*, are unparalleled.

On July 15, 1941, the head and *Oberster Befehlshaber* of the *Wehrmacht* awarded

Kommodore Oberst Mölders, after his 101st air victory in the *Freiheitskampf* of the German people, the highest medal of bravery: the *Eichenlaub mit Schwertern und Brillanten zum Ritterkreuz des Eisernen Kreuzes.*

In recognition of Mölders' services the head and *Oberster Befehlshaber* of the *Wehrmacht* ordered that the victorious *Jagdgeschwader*, led by Werner Mölders until now, carry his name in the future. At the same time the *Führer* was arranging a state funeral for Mölders.

April 7, 1942
OKW-Bericht: The *Jagdgeschwader* "Mölders" achieved its 3000th air victory yesterday.

August 2, 1942
OKW-Bericht: During heavy air battles that developed yesterday in the middle front sections, the *Jagdgeschwader* "Mölders" shot down 25 Soviet planes, despite unfavorable weather conditions.

July 12. 1943
Press report: On 10 July the *Jagdgeschwader* "Mölders," led by *Oberstleutnant* Nordmann, reported their 5500th shootdown after the members of this *Geschwader* shot down the 5000th enemy plane on 2 June.

July 29, 1943
Press report: The *Jagdgeschwader* "Mölders," led by the bearer of the *Eichenlaub*, *Oberstleutnant* Nordmann, achieved its 6000th shootdown on the Eastern Front.

September 16, 1943
OKW-Bericht: The *Jagdgeschwader* "Mölders" achieved its 7,000th shootdown on 15 September.

May 4, 1944
OKW-Bericht: The standing *Jagdgeschwader* "Mölders," under the leadership of *Oberstleutnant* Nordmann, reported its 8,000th air victory.

List of Werner Mölders' Air Victories

lfd. No.	Date	Type	Location
Spain			
1	7-15-1938	Curtiss	Algar region
2	7-17-1938	Curtiss	Region north of Liria
3	7-19-1938	Rata	Region west of Villar del Arzobispo
4	8-19-1938	Rata	Flix region
5	8-23-1938	SB-2	Albi region
6	9-9-1938	Rata	Air above Flix region
7	9-13-1938	Rata	Air above Flix region
8	9-23-1938	Rata	Region southeast of Ginestar
	9-23-1938	Rata	Shootdown not confirmed
9	10-10-1938	Rata	Region northeast of Flix
10	10-15-1938	Rata	Region west of La Figuera
11	10-15-1938	Rata	Sierra de Montsant region
12	10-31-1938	Rata	Region northwest of Flix
13	10-31-1938	Rata	Region south of Ribarroja
14	11-3-1938	Rata	Mola region
Western Front			
1	9-20-1939	Curtiss	by Sierck
2	10-30-1939	Blenheim	by Klüsserath/Mosel
3	12-22-1939	Hurricane	15 km NE Metz
4	3-2-1940	Hurricane	by Völklingen
5	3-3-1940	Morane	by Metz
6	3-26-1940	Morane	by Diedenhofen
7	4-2-1940	Hurricane	S Saargemünd
8	4-20-1940	Curtiss	S Niedergailbach
9	4-23-1940	Hurricane	S Diedenhofen
10	5-14-1940	Hurricane	Sedan-Charleville region
11	5-15-1940	Hurricane	Sedan region
12	5-19-1940	Curtiss	NE Reims
13	5-20-1940	Wellesley	by Compiégne
14	5-21-1940	Morane	
15	5-21-1940	Morane	
16	5-21-1940	Morane	
17	5-22-1940	Potez 63	SW Mourmelon Airport
18	5-25-1940	Morane	Wald Villers Cotterets
19	5-27-1940	Curtiss	15 km SW Amiens
20	5-27-1940	Curtiss	15 km SW Amiens
21	5-31-1940	LeO 45	Abbéville-Amiens region
22	6-3-1940	Spitfire	Above Paris
23	6-3-1940	Curtiss	Above Paris
24	6-5-1940	Bloch	Beauvais-Compiégne region
25	6-5-1940	Potez 63	Beauvais-Compiégne region
Channel Front			
26	7-28-1940	Spitfire	Above Dover
27	8-26-1940	Spitfire	Folkestone
28	8-28-1940	Curtiss	NE Dover
29	8-28-1940	Hurricane	Canterbury
30	8-31-1940	Hurricane	Between Folkestone and Dover

German Fighter Ace Werner Mölders - An Illustrated Biography

31	8-31-1940	Hurricane	Between Folkestone and Dover
32	8-31-1940	Hurricane	Between Folkestone and Dover
33	9-6-1940	Spitfire	by Folkestone
34	9-7-1940	Spitfire	by London
35	9-9-1940	Spitfire	by London
36	9-11-1940	Hurricane	SE London
37	9-14-1940	Spitfire	SW London
38	9-16-1940	Hurricane	S London
39	9-20-1940	Spitfire	by Dungeness
40	9-20-1940	Spitfire	by Dungeness
41	9-27-1940	Spitfire	by Maidstone
42	9-28-1940	Spitfire	by Littlestone
43	10-11-1940	Spitfire	by Folkestone
44	10-12-1940	Hurricane	Between Lympne and Canterbury
45	10-12-1940	Hurricane	Between Lympne and Canterbury
46	10-12-1940	Hurricane	by Dungeness
47	10-15-1940	Hurricane	by Kenley
48	10-17-1940	Spitfire	by London
49	10-22-1940	Hurricane	NW Maidstone
50	10-22-1940	Hurricane	NW Maidstone
51	10-22-1940	Hurricane	NW Maidstone
52	10-25-1940	Spitfire	NW Dover
53	10-25-1940	Spitfire	by Margate
54	10-29-1940	Hurricane	by Dungeness
55	12-1-1940	Hurricane	by Ashford
56	2-10-1941	Hurricane	5 km NNE Calais
57	2-20-1941	Spitfire	by Dover
58	2-20-1941	Spitfire	by Dover
59	2-25-1941	Spitfire	N Gravelines
60	2-26-1941	Spitfire	SE Dungeness
61	3-12-1941	Spitfire	by Dungeness
62	3-13-1941	Spitfire	SW Boulogne
63	4-15-1941	Spitfire	by Boulogne
64	4-16-1941	Hurricane	SW Dungeness
65	4-16-1941	Spitfire	W Berck
66	5-4-1941	Hurricane	E Deal
67	5-6-1941	Hurricane	by Dover
68	5-8-1941	Spitfire	Outside Dover into water

Eastern Front

69	6-22-1941	Curtiss	
70	6-22-1941	SB-2	
71	6-22-1941	SB-2	
72	6-22-1941	SB-2	
73	6-24-1941	SB-2	
74	6-25-1941	SB-2	
75	6-25-1941	SB-2	
76	6-29-1941	Pe-2	
77	6-29-1941	I-16	
78	6-30-1941		
79	6-30-1941		
80	6-30-1941		
81	6-30-1941		
82	6-30-1941		
83	7-5-1941	SB-2	
84	7-5-1941	SB-2	
85	7-5-1941	I-18	
86	7-5-1941	I-18	
87	7-9-1941	Curtiss	
88	7-9-1941	Curtiss	
89	7-9-1941	I-16	
90	7-10-1941	RZ	
91	7-10-1941	RZ	
92	7-11-1941		
93	7-11-1941		
94	7-12-1941		
95	7-13-1941		
96	7-13-1941		
97	7-14-1941	Pe-2	
98	7-14-1941	Pe-2	
99	7-14-1941	Pe-2	
100	7-15-1941		
101	7-15-1941		

Duplicate:

Mölders, *Hauptmann*
Name,   Rank

Wiesbaden, September 20, 1939
Location, Date

Air Combat Report

1. <u>Time (day, hour, minute) and region</u>:   September 20, 1939 14:50 west of Merzig over French territory

2. <u>Type of plane shot down</u>: Curtiss

3. <u>Nationality of the opponent</u>: French

4. <u>Type of destruction</u>: fire on impact
    a) burning
    b) dismantled
    c) forced landing, where

5. **Kind of impact**: on the other side
    a) this side
    b) other side

6. <u>Fate of the passengers of the enemy plane</u>: parachute jump

7. <u>Combat report</u>: I took off with my *Schwarm* at 14:27 towards 6 reported enemy monoplanes south of Trier. At 4500 m the *Schwarm* flew over the Saar near Merzig as 6 planes were spotted at an altitude of 5000 m south of Conz. I crossed the opponent in a circle to the north and proceeded with a surprise attack on the back of the plane. From 50 m I opened fire, from which the Curtiss began to swim. After another longer burst of fire, smoke came from the plane and parts flew off. It tipped forward, and I lost it because I had to ward off a newly appeared opponent.

8. <u>Witness of the shootdown</u>:
    a) <u>air</u>: *Lt.* Brandhuber, *Uffz.* Bleidorn, *Uffz.* Freund
    b) ground: ?

9. <u>Witness report</u>:

Werner Mölders
Pictures and Documents

German Fighter Ace Werner Mölders - An Illustrated Biography

Childhood and School Days

Right: Mölders was born in this house in Gelsenkirchen/Westfalen. He spent his first years here until the move to Brandenburg an der Havel.

Bottom left: Werner (left) with his brother, Hans, and his sister, Annemarie.

Right: Mother Anna-Maria Mölders with her children Werner, Victor, and Hans (left to right).

German Fighter Ace Werner Mölders - An Illustrated Biography

Mölders spent his youth in Brandenburg an der Havel. View of the Rathenower Torturm, a famous landmark of the city.

Bottom left: This picture on Werner Mölders' school identity card was taken in June 1925 in Brandenburg/Havel.

Right: This picture was taken of Werner with his siblings, Annemarie and Victor, around the same time.

45

German Fighter Ace Werner Mölders - An Illustrated Biography

Summer 1936 on the Havel: Werner (kneeling, to the right) with his brother, Victor (laying in front of him), as well as his cousins. Bottom: Vacation with his beloved Uncle Paul (in the background) in Dahlen/Mark Brandenburg. Werner and Victor with their childhood friend, Ruth Fuhrmeister, shortly before Werner "went hunting," equipped with a *Luftgewehr* and accompanied by his "hunting dogs."

German Fighter Ace Werner Mölders - An Illustrated Biography

After flying for the first time at 11 years old on a sightseeing flight above Trier, he never let go of his dream of becoming a pilot. Mölders took advantage of every opportunity to fly, that is, to realize his dream. Here he is boarding a *Verkehrsmaschine* of the German *Lufthansa* with his youth group in Berlin-Tempelhof.

German Fighter Ace Werner Mölders - An Illustrated Biography

As *Oberprimaner* in the official "*Studentenkluft*" in front of a "Brennabor," one of the better automobiles of this time. The picture was taken in the spring of 1928 in Dahlen on the occasion of his Uncle Paul's visit.

He has done it! After the successful graduating students of the "Saldria-*Gymnasium*" in Brandenburg/Havel received their school-leaving certificates in the spring of 1931, they posed for the photographer. To the far left in the last row, Werner Mölders.

Military and Aviation Training

Shortly before his school-leaving exam, Mölders took a difficult entrance exam for entry as *Offiziersanwärter* in the *Reichswehr*'s 100,000-*Mann-Heer*. Out of 60 applicants only three were accepted, and one of them was Mölders!

After his recruit training with the 7. *Kompanie* (II.Gren.Batl.) of the Inf.Regt.2 in Allenstein/East Prussia. Mölders returned to Rastenburg, where he reported for duty. The picture shows him (standing) during the farewell party at the end of September 1931 with comrades and his instructor, *Korporal* (*Unteroffizier*) Schimanke (second from left).

As *Obergefreite*, Mölders (sitting, second from the right) trained the recruits of the 2. *Abteilung* of the 15./*Inf.Regt*.2 in winter 1931/32.

German Fighter Ace Werner Mölders - An Illustrated Biography

As of October 1932 Mölders attended the *Kriegsschule* in Dresden. He recuperated from the difficult and extensive training on a ski trip in Czechoslovakia in March 1933 that ended with broken ski tips.

Left: snapshot of the *Fähnrichs* course: Mölders with his comrade, Klaus Nöske (Nöske was awarded the *Ritterkreuz* on May 16, 1941, as *Hptm.* and *St.Kpt.* of 1./KG 4 "General Wever."). Right: Mölders during an excursion on top of the "Dybin" in Czechoslovakia.

Ib Lehrgang   Dresden, den 10. Juni 1933.
Infanterieschule.

Abgangszeugnis
des
Ib Lehrgangs der Infanterieschule

Der Fahnenjunker-Unteroffizier   M ö l d e r s
Pionier Bataillon 1   (Infanterie Regiment 2)

hat in der im April 1933 an der Infanterieschule abgehaltenen Fähnrichsprüfung

1. in der Taktik	ziemlich gute
2. in der Geländekunde	ziemlich gute
3. im Zeichnen	gute
4. in der Waffenlehre	ziemlich gute
5. in der Pionierlehre	ziemlich gute
6. in Luftschutz und Tarnung	genügende
7. im Nachrichtenwesen	genügende
8. im Kraftfahrwesen	fast genügende
9. im Heerwesen	ziemlich gute
10. in der Staatsbürgerkunde	ziemlich gute
11. in Fremdsprachen	genügende
12. im Waffendienst	gute
13. in den Leibesübungen	gute
14. im Reiten	ziemlich gute

Kenntnisse bewiesen und im ganzen   ziemlich gute   Leistungen gezeigt.

Oberstltnt. und Lehrgangsleiter an der Infanterieschule.

The certificate for the Fähnrich's test that Mölders passed with "rather great" effort.

Ib Lehrgang   Dresden, den 9. Juni 1933.
Infanterieschule

B e s c h e i n i g u n g.

Der Fahnenjunker-Unteroffizier ..M.ö.l.d.e.r.s..........
........................Pionier.Bataillon.1...... hat die
Fähnrichsprüfung 1933 mit ..ziemlich.gut........bestanden.
Er ist gemäß Verfg.Rw.Min ... v. ..6.6.33..
mit Wirkung vom .... zum Fähnrich be-
fördert.

Oberstleutnant u. Lehrgangsleiter.

As a result of June 1, 1933, he was promoted to Fähnrich.

German Fighter Ace Werner Mölders - An Illustrated Biography

The fledgling *Fähnrich* was transferred in June 1933 to 1./ *Pionier Bataillon* 1, and came to the *Pionierschule* in München. Below: Mölders, second from the left, with his comrades in München-Schleiflheim.

The *Fähnriche* took part in the "*Tag der Pioniere*" in July 1933 in Ingolstadt/Donau. The *Pontonbrücke* that was built within a short time above the Donau.

A rare shot! The *Fähnriche* during *Kraftfahrdienst*. On motorcycles, from left to right: Strümpell, Mölders, Nöske, and *Leutnant* Gang.

German Fighter Ace Werner Mölders - An Illustrated Biography

His time at the *Pionierschule* in München was one of Mölders' most beautiful experiences. He spent his free time as a water sports enthusiast on the Starnberger See (see top picture), and in the Bavarian mountains.

Mölders (to the far right) drinks a glass of milk during a rest after a mountain hike on the "First-Alm" above the Spitzingsee.

After the "defeat" of the "Alpspitze," a rest on the mountaintop with a well deserved snack (Mölders, second from the right).

Vacation at home. Victor and Werner lift up their mother on the roof terrace of their home.

The "Mölders boys" are all at home again. From left to right: Hans, Victor, and Werner.

German Fighter Ace Werner Mölders - An Illustrated Biography

In winter 1933/34, training at the *Pionierschule* in München continued. As a *Zugführer*, Mölders drove in an open PKW with his men despite the snow and cold.

Along with automobiles and motorcycles, the horses were constantly moved around on the ground. Mölders—in the foreground—with his comrades shortly before the ride to practice. Left Gerhard Kollewe, who was awarded with the *Eichenlaub zum Ritterkreuz des Eisernen Kreuzes* on August 12, 1942. Unfortunately, he fell as *Major* and *Kdr.* of II./KLG 1 during air battle with a Spitfire above the Mediterranean Sea on October 17, 1942.

Kommando
der Pionierschule.

München, den 25. Januar 1934.

Ausweis.

Der Fähnrich *Mölders* des *1.* Pion.Batl. ist mit Funkspruch des Reichswehrministeriums vom 25.Jan.1934 zum Oberfähnrich befördert.

Oberstleutnant und Kommandeur.

His time at the *Pionierschule* in München ended with his promotion to *Oberfähnrich* (see picture), and his commitment to service in the *Reichsheer*.

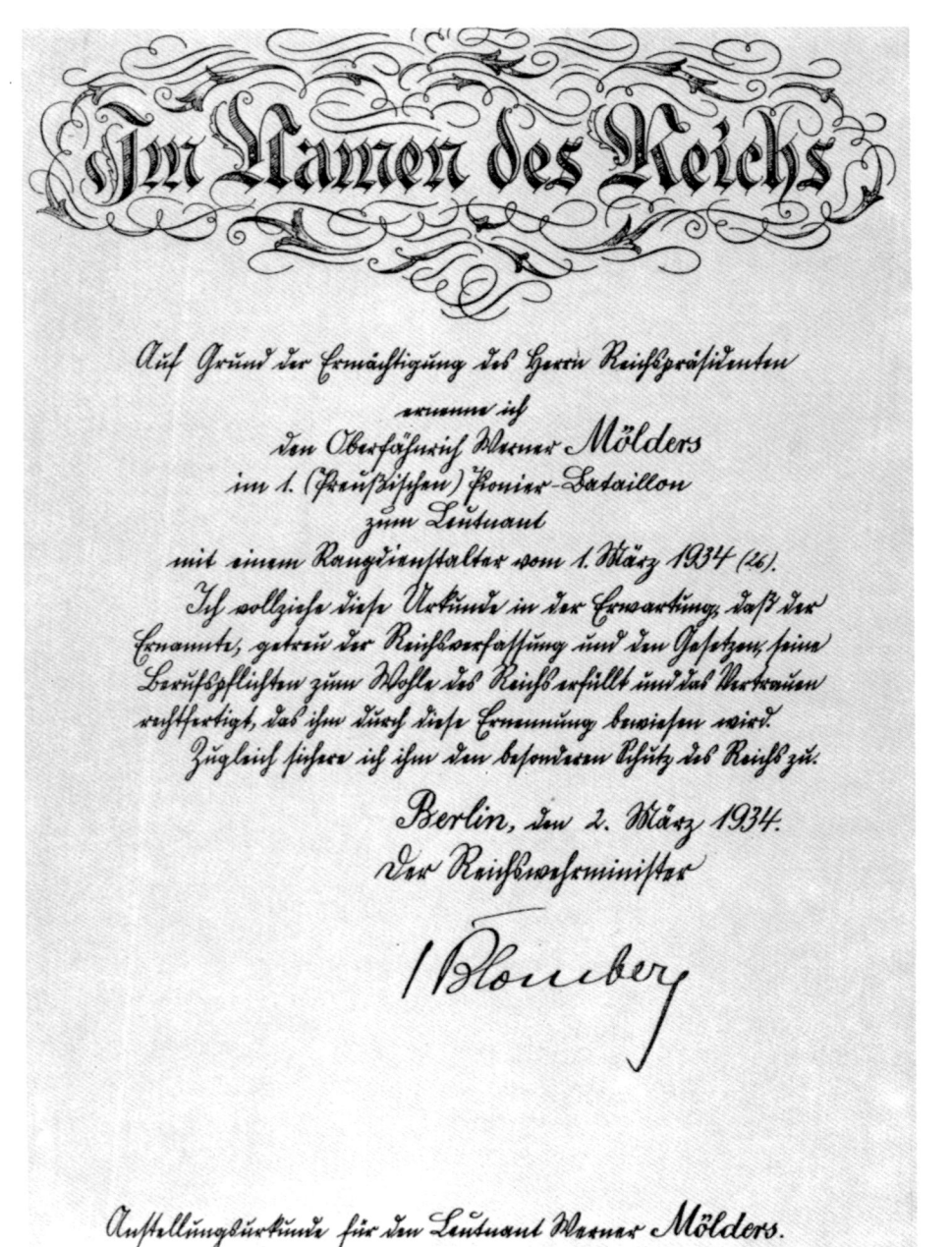

Im Namen des Reichs

Auf Grund der Ermächtigung des Herrn Reichspräsidenten ernenne ich den Oberfähnrich Werner Mölders im 1. (Preußischen) Pionier-Bataillon zum Leutnant mit einem Rangdienstalter vom 1. März 1934 (26).

Ich vollziehe diese Urkunde in der Erwartung, daß der Ernannte, getreu der Reichsverfassung und den Gesetzen, seine Berufspflichten zum Wohle des Reichs erfüllt und das Vertrauen rechtfertigt, das ihm durch diese Ernennung bewiesen wird. Zugleich sichere ich ihm den besonderen Schutz des Reichs zu.

Berlin, den 2. März 1934.

Der Reichswehrminister

v. Blomberg

Aushändigungsurkunde für den Leutnant Werner Mölders.

The letter of appointment to *Leutnant*, signed by *Reichswehrminister Generaloberst* von Blomberg.

View of Cottbus, for the pilots an unforgettable, charming city in Mark Brandenburg, in which the *Deutsche Verkehrsfliegerschule* found its home.

Werner Mölders also learned to fly at the *Deutsche Verkehrsfliegerschule* in Cottbus, and officially belonged to it as a flight student and pilot from February 6 to December 31, 1941. The picture shows him in front of a Focke-Wulf AL 101 D (Argus As 8/A) of DVS Cottbus (D-2818).

German Fighter Ace Werner Mölders - An Illustrated Biography

In the DVS many types of planes were flown, including the Al 102 (top picture) and the He 45 (center picture). The picture at the bottom shows that there were accidents. During landing this He 45 went head first onto the snow-covered runway; a typical "pilot memorial."

German Fighter Ace Werner Mölders - An Illustrated Biography

Left: Werner Mölders (center) with his schoolfellows and comrades, *Ltn*. Rosenthal (right) and *Ltn*. Darjes, who was awarded with the *Ritterkreuz* as *Major* and *Kdr*. of II./Sch.G. 1 on October 14, 1942. Right: The young *Ltn*. Mölders presents himself proudly here in the uniform of the *Deutscher Luftsport-Verband* (DLV) in spring 1934.

A *Kette* Arado Ar 66, with which the flight students gladly flew.

61

German Fighter Ace Werner Mölders - An Illustrated Biography

Above: The elegant and outstanding Fw 56 was also a favorite plane. Below: View of the air traffic of the pilots in training.

With an Arado Ar 66 Mölders carried out a high-altitude flight on August 9, 1934, as the "Barogramm" of the DVS Cottbus below proves.

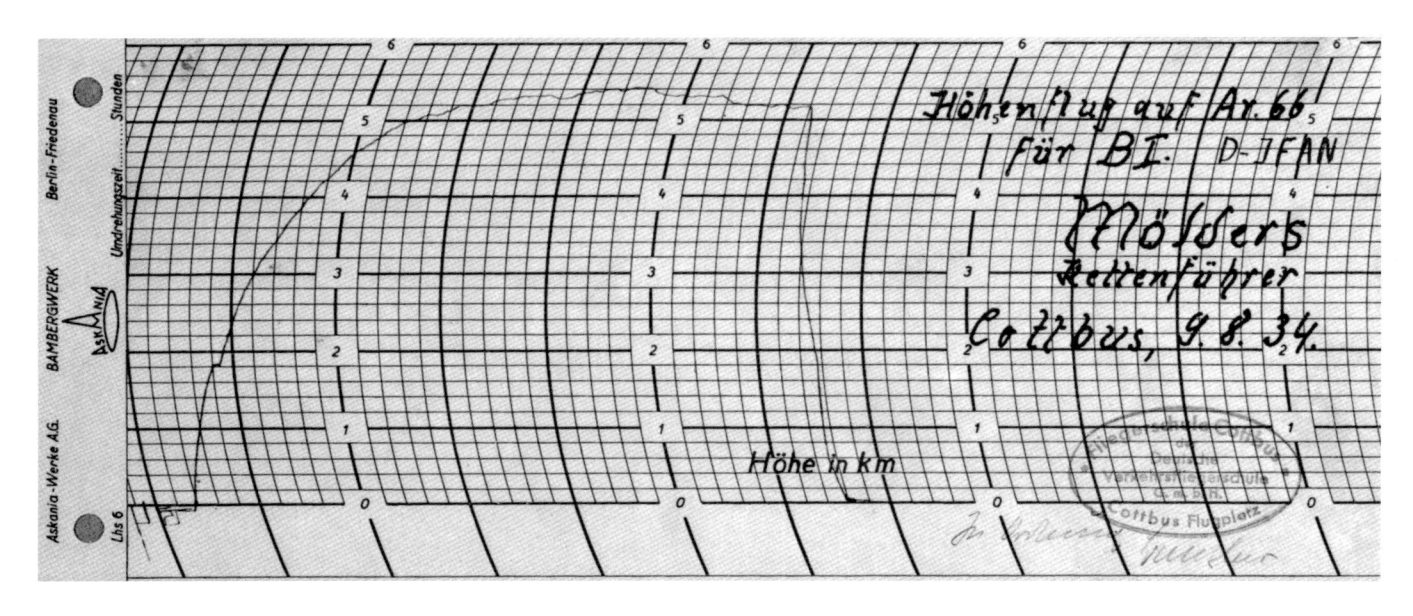

German Fighter Ace Werner Mölders - An Illustrated Biography

Left: From January 1 to June 30, 1935, Mölders was transferred as a pilot with four of his comrades to the *Kampffliegerschule* in Tutow. The five *Leutnants* from left to right: Frhr. von Beust, Keil, Roth, Mölders, and Mund. It is clearly shown that two of the *Offiziere* still have the symbol of the DLV on their tunics, while the others already wear the new national emblem of the *Luftwaffe*, the eagle with the swastika.

Center: At this *Kampffliegerschule* the Dornier Do 23, a bomber plane that was very difficult to fly, was also flown.

Below: A very elegant plane, however, was the Heinkel He 70 "Blitz," one of which is being serviced here.

German Fighter Ace Werner Mölders - An Illustrated Biography

Mölders ordered a short vacation on Easter in the beloved Bavarian mountains with comrades during three month fighter pilot training in München-Schleiflheim. The picture shows him and his friends during a rest in the morning sun on the "Vorderbrand-Alm."

Mölders and von Harling on a boat ride on the Königssee by Berchtesgaden, and on a small hike with his friends in "Maierwinkel" with a view of the lake.

With this certificate Mölders was awarded with the new *Flugzeugführerabzeichen* of the *Luftwaffe* on May 21, 1935. He reached his goal of becoming a pilot, despite all initial difficulties.

Mölders took off for a long distance flight with the Junkers W 34 (with BMW "Hornet") from München-Schleiflheim.

German Fighter Ace Werner Mölders - An Illustrated Biography

During a flight above the Thüringer forest.

Above the center of München.

German Fighter Ace Werner Mölders - An Illustrated Biography

For his shining theatrical efforts as Ikaros in the first and only fighter pilot opera, "Ikarus and Ikarine," at the DVS in Cottbus on October 10, 1934, the *Staatssekretär* Milch presented Mölders and Ltn. Petersson each a free ticket to Greece from Lufthansa. Mölders (in the foreground) as "Ikarus" in the third act.

Center: On June 6, 1935, both friends redeemed their free tickets and flew Lufthansa with a Ju 52 to the sunny south, where they gladly enjoyed the beauties of Greece.

Below: This souvenir photograph shows Mölders at the Acropolis.

German Fighter Ace Werner Mölders - An Illustrated Biography

At the course in München-Schleiflheim that, among other things, was used as preparation for missions with a newly formed *Stuka-Gruppe*, they also flew the Arado Ar 65, with which diving was practiced from high altitudes.

From July 1935 until March 1936 Mölders belonged to the *Fliegergruppe* ("Immelmann") in Schwerin. The picture shows the beaming, young *Leutnant* at the entrance to the air base, wearing the *Flugzeugführerabzeichen* on his chest.

German Fighter Ace Werner Mölders - An Illustrated Biography

His own PKW, an "Adler," can't be missed!

The time in Schwerin brought the young *Offiziere* a beautiful and almost carefree life that they always dreamt of. They were often guests in the "Villa BALDAUF," where Mölders also met his later wife. Right: (from left to right): Mölders, Luise Baldauf, Balfanz, and Keil. Above: Mölders, Baldauf, Luise Baldauf, Keil, and Krause.

As a Fighter Pilot in the new German *Luftwaffe*

Left: After the lifting of the demilitarized zone in the Rheinland, *Jagdstaffeln* of the new German *Luftwaffe* flew over the Rhine on March 7, 1936, for the first time. Mölders flew from Lippstadt/Westf. with his *Staffel* over the Ruhrgebiet on the Rhine. They landed in Düsseldorf, and were enthusiastically greeted by the people.

From April 1, 1936, until March 14, 1937, Mölders belonged to the II. Gruppe of *Jagdgeschwader* 134 "Horst Wessel" in Werl/Westfalen. Mölders (right) with the *Kommandeur* of the *Gruppe*, *Major* Osterkamp, and *Obltn*. Kienzle on April 7, 1936, shortly before the parade on the occasion of the move to the new base in Werl.

With drawn swords Osterkamp marches his men in "great service dress." Beside them *Hptm*. Vollbracht (in 1945 *Oberst*, and bearer of the *Ritterkreuz*) and *Obltn*. Kienzle. Left at the edge of the picture greeting *Ltn*. Mölders, *Ordonnanzoffizier* of the *General*, who takes off the parade. Osterkamp, known by all pilots as "Uncle Theo," was the bearer of the "Pour le mérite" (32 air victories in WWI) and of the *Ritterkreuz* (conferment on August 22, 1940).

> **Im Namen des Reichs**
>
> ernenne ich den Leutnant
>   Werner Mölders
> in der Fliegergruppe II./134 Werl
> mit Wirkung vom 1. April 1936 zum
>   Oberleutnant
> mit einem Rangdienstalter vom 1. April 1936 (6).
>
> Ich vollziehe diese Urkunde in der Erwartung, daß der Ernannte getreu seinem Diensteide, seine Berufspflichten gewissenhaft erfüllt und das Vertrauen rechtfertigt, das ihm durch diese Ernennung bewiesen wird.
> Zugleich darf er des besonderen Schutzes des Führers und Reichskanzlers sicher sein.
>
>   Berlin, den 20. April 1936.
>   Namens des Führers und Reichskanzlers
>   Der Reichsminister der Luftfahrt
>   und Oberbefehlshaber der Luftwaffe.
>
>   *Hermann Göring*

On April 20, 1936, Mölders was promoted to *Oberleutnant* as a result of April 1, 1936.

The first picture as *Oberleutnant* with the second wing on his collar.

Mölders as head of 1. *Jagdschlstaffel* of the *Geschwader* in Werl. In the picture above, to the left of him, and in the picture below, marching to the side, is *Ltn*. Hachfeld. Unfortunately this outstanding pilot fell to his death as bearer of the *Ritterkreuz*, *Hptm*. and *Kdr*. of III./ZG 2. on December 2, 1942, during takeoff in Bizerta/Tunesien after 650 missions, 230 of which were *Jabo* missions.

Air drills with the Arado Ar 68: Flight in *Staffelwinkel* (above) and *Staffelkeil* (below).

German Fighter Ace Werner Mölders - An Illustrated Biography

View from the pilot's seat of the Arado Ar 68 of the diving comrades.

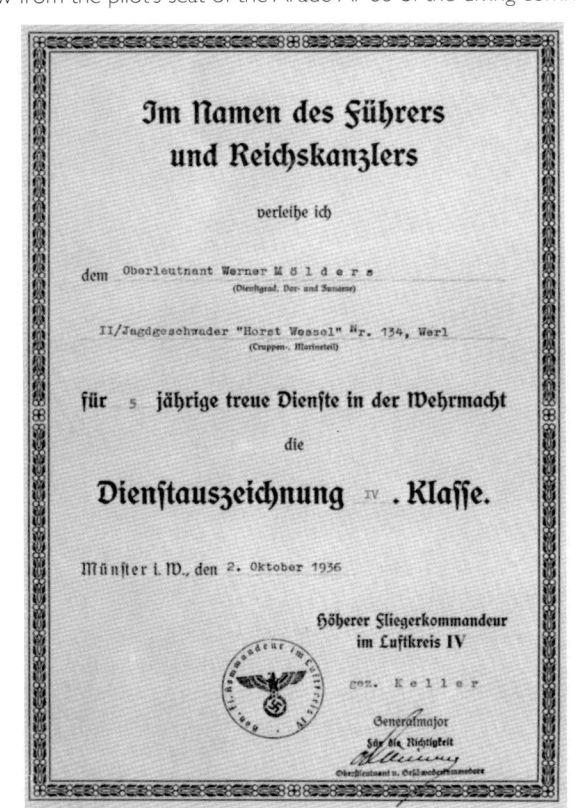

On October 2, 1936, Mölders received this certificate and the *Dienstauszeichnung IV. Klasse* for his five year membership in the *Wehrmacht* (before *Reichsheer*).

German Fighter Ace Werner Mölders - An Illustrated Biography

These pictures were taken during a visit of the Bad Lippspringe airport in 1937. Above, 6. *Staffel* of II./JG 234 with its Arado Ar 68-E; Center, Heinkel-*Jagdflugzeuge* He 51 during a landing approach; below: a *Kampfflugzeug* Junkers Ju 86.

German Fighter Ace Werner Mölders - An Illustrated Biography

Top left: From March 15, 1937, until April 13, 1938, Mölders was *Kapitän* of 1. *Staffel* in I./JG 334 (later 1./JG 53 "Pik As") in Wiesbaden. Mölders (third from the right) with his pilots shortly before takeoff with their He 51. Second from the right, Heinz Bretnütz, later a "*Jagdflieger-As*," as well.

Right: A member of the ground crew as a sentry with a flare pistol. In the background a mechanic is servicing one of the new Messerschmittjäger Bf 109s.

Center: With the new planes flying is twice the fun! A Bf 109 B *Schwarm* during a flyover of the area in a demonstration.

The parked, new Bf 109 B planes are the pride and joy of the German *Jagdwaffe*.

German Fighter Ace Werner Mölders - An Illustrated Biography

With the "Legion Condor" in Spain

On April 14, 1938, Mölders was ordered as *Staffelkapitän* to 3./J 88 of the "Legion Condor" in Spain. The picture shows him in the uniform of the legion shortly before his arrival to the unit.

The *Staffel* was equipped with the Heinkel He 51 A-2, and made themselves a name under the leadership of *Obltn.* Galland during many successful low-flight attacks to support the ground troops above the borders of the legion.

Bottom left: Galland, kneeling at the *Maschinengewehr*, and his men eagerly, and somewhat skeptically, observe the landing of a comrade.

Right: On May 24, 1938, Mölders officially took over Galland's *Staffel*. Galland (left) shortly before the flight home, already in civilian clothes but still wearing the "*Schifferl-Mütze*" of the legion on his head. To the right, wearing a hat, "Edu" Neumann, later one of the most popular fighter pilots of the *Luftwaffe* (among others, *Kommodore* of JG 27 in Africa and the *Reichsverteidigung*).

German Fighter Ace Werner Mölders - An Illustrated Biography

Mölders continued to fly successful ground attack flights on the Southern Front with his He 51 *Staffel*. During an exchange of views, from left to right: *Ltn.* Lippert, Mölders, *Ltn.* Ströflner, and *Obltn.* Reents.

During this time Mölders undertook many trips to the Front and visited the front lines. Here in the trenches (right) in conversation with Spanish national soldiers.

Mölders remained in close contact with the *Führungsstab* of the Legion. The picture shows the *Stab* on top of a hill near Villar. Right, with binoculars, *Generalmajor* Frhr. von Richthofen, the last *Kdr.* of the Legion, who in 1936/37 was already the head of the *Stab* in Spain.

German Fighter Ace Werner Mölders - An Illustrated Biography

At the beginning of July 1938 3. *Staffel* was finally equipped with the modern and fast Messerschmitt Bf 109, after the 1. and 2. *Staffel*. The joy was great as the first planes landed.

View of the pilot's seat in the Bf 109 B-2.

View of the Bf 109 B-2 motor with trunk armament, 2 x MG 17.

German Fighter Ace Werner Mölders - An Illustrated Biography

The shootdown of an I-15 ("Curtiss") on July 17, 1938—top picture a plane of this type—was confirmed as Mölders' first official air victory, as this certificate proves. In reality it was already his second shootdown, because two days before he defeated his second opponent in the Algar region in an air combat. This shootdown was, according to the certificate, recognized as his third air victory. Every fighter pilot who achieved shootdowns with the "Legion Condor" received these certificates.

German Fighter Ace Werner Mölders - An Illustrated Biography

Top left: View of La Cenia air field, that was occupied by 3./J 88 for a long time. The city is situated on a river with the same name not far from the mouth of the river.

Next to it: Mölders and his men lived in this nicely situated house on the border of the city for almost seven months.

A cheerful morning table in their quarters.

A game of chess during mission breaks on the veranda of the country house. Fözö (left) is playing against Mölders.

German Fighter Ace Werner Mölders - An Illustrated Biography

Mölders and his *Staffel* achieved success after success, which received great recognition. On October 18, 1938, he was promoted to Hauptmann at 25 years of age. This picture shows him shortly after the promotion.

Im Namen
des
Führers und Reichskanzlers
befördere ich

den Oberleutnant in der Luftwaffe
Werner Mölders
mit Wirkung vom 1.Oktober 1938 zum
Hauptmann

Ich vollziehe diese Urkunde in der Erwartung, daß der Genannte getreu seinem Diensteide seine Berufspflichten gewissenhaft erfüllt und das Vertrauen rechtfertigt, das ihm durch diese Beförderung bewiesen wird. Zugleich darf er des besonderen Schutzes des Führers und Reichskanzlers sicher sein.

Berlin, den 18. Oktober 1938

Der Reichsminister der Luftfahrt
und Oberbefehlshaber der Luftwaffe

Göring

The fighter pilots' every day life on the air field in Spain. The "men in black" of the ground crew did a good job under pressure, and contributed to the success of their pilots.

German Fighter Ace Werner Mölders - An Illustrated Biography

With 12 acknowledged air victories, *Hptm*. Schellmann was the most successful fighter pilot of the legion in Spain after Mölders. As a bearer of the *Ritterkreuz* and *Kommodore* of JG 27 he had to jump with a parachute on June 22, 1941, near Grodno (Russia), and has been missing ever since.

Bottom left: A Bf 109 *Schwarm* on a mission over the Sierra. On the rudder assembly of the plane in the foreground five lines marking shootdowns can be seen.

Right: Morning mission conference of 3./J 88. Left Mölders, with the map in his hand, picture center, with a map as well, "Joschko" Fözö. Fözö flew 147 missions in Span and won 3 air victories (was awarded with the *Ritterkreuz* on July 1, 1941).

On June 6, 1939, the shootdown of a Martin-Bomber was recognized as his 14th and last air victory (it was actually his 5th shootdown). Top right: Werner Mölders, the most successful fighter pilot of the "Legion Condor."

German Fighter Ace Werner Mölders - An Illustrated Biography

Above: The mechanics during the last preparations before takeoff. In the foreground the "heavy haul" of the *Kampffliegerstaffel* of the Legion (K/88) in the area, as well.

Center: A Bf 109 of 3./J 88 flies over the area in a low-level flight.

Left: On December 5, 1938, Mölders flew in a Ju 52 from Platz La Cenia to his home. His Bf 109 stayed back, and the rudder assembly was marked with 15 air victories (14 were acknowledged).

German Fighter Ace Werner Mölders - An Illustrated Biography

June 6, 1939: Reception of the "Legion Condor" in Berlin. Left, as *Fahnenbegleiter* of the pennants, conferred by Franco, of the Legion Wolf—Dietrich Wilcke, who, unfortunately, fell in March 1944 as one of the best German fighter pilots in the "*Reichsverteidigung*."

Hitler inspects the troops lined up for the parade. From left to right: Warlimont, Volkmann, Sperrle, and v. Richthofen were the former commanders of the legion in Spain.

German Fighter Ace Werner Mölders - An Illustrated Biography

Tribute to the fallen combatants in Spain on the pleasure grounds in Berlin. Each white plaque carried the name of one of the fallen.

Mölders was awarded with the "*Deutsches Spanien-Kreuz in Gold mit Schwertern und Brillanten*," as well as two Spanish medals for his outstanding efforts during the mission in Spain. The conferment certificates are illustrated here.

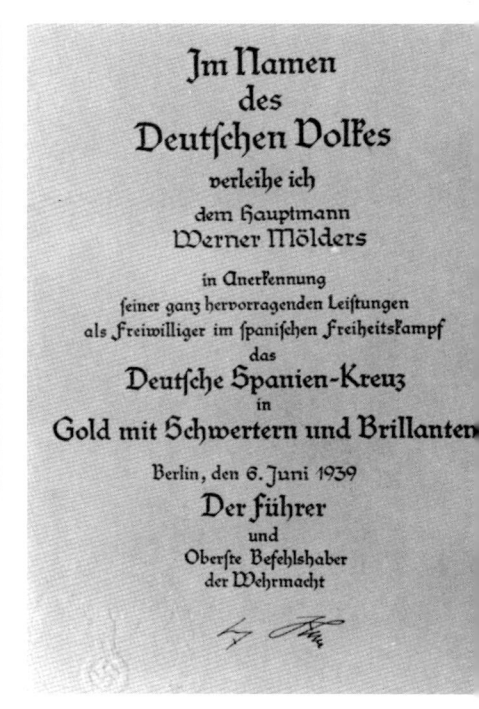

MERKBLATT
FÜR DEN 6. JUNI 1939

1. Antreten und Meldung in der Marmorgallerie.
   Mütze aufbehalten. Koppel umgeschnallt lassen.
   Aufstellung in Linie zu 2 Gliedern.
   Am rechten Flügel die 3 Befehlshaber der Legion Condor, Admiral Carls, Admiral Boehm Oberst d. G. Warlimont und die Offiziere des Sonderstabs W.

   Dann nebeneinander
   die spanischen Offiziere
   die italienischen Offiziere
   Sonderstab W
   die Kommandeure und Kommandanten
   die Offiziere mit Spanienkreuz in Gold mit Brillanten
   die Offiziere mit Spanienkreuz in Gold
   die Unteroffiziere mit Spanienkreuz in Gold.

   Innerhalb der einzelnen Blocks nach den Wehrmachtteilen antreten.
   Reihenfolge: Luftwaffe, Heer, Kriegsmarine.
   Innerhalb der Wehrmachtteile dem Dienstalter nach aufstellen.

2. Der Führer betritt die Gallerie von der Mitte der Nordseite der Gallerie.

3. General der Flieger Sperrle meldet dem Führer. Dabei grüßen alle Generale und Admirale, während alle anderen Offiziere und Unteroffiziere in Grundstellung bleiben und Augen rechts bzw. links nehmen.

4. Der Führer geht die Front vom rechten Flügel aus ab. Er wird begleitet von den Oberbefehlshabern der 3 Wehrmachtteile, Generaloberst Keitel und Generaloberst Milch. Die Befehlshaber der Legion Condor und Oberst d. G. Warlimont schließen sich an.

5. Nachdem der Führer die Front abgeschritten hat, begibt er sich in sein Arbeitszimmer. Es folgen ihm alle deutschen und ausländischen Generale und Admirale. Die anderen angetretenen Offiziere und Unteroffiziere begeben sich sofort in den Mosaiksaal und Speisesaal und verteilen sich dort auf die Tische. Die auf der Rückseite namentlich aufgeführten Offiziere und Unteroffiziere nehmen an den numerierten Tischen Platz. Es bleiben alle Herren an ihren Stühlen stehen bis der Führer den Marmorsaal betreten und sich gesetzt hat.

6. Der Führer wird zuerst an einem Tisch in dem Mosaiksaal, dann an einem Tisch im Speisesaal Platz nehmen. Wenn er den Speisesaal betritt, haben sich alle Herren von ihren Plätzen zu erheben.

7. Nach dem Essen kann zwanglos herumgegangen werden.

8. Das Rauchen ist in sämtlichen Räumen der Reichskanzlei nicht gestattet, auch nicht im Ehrenhof und Garten.

The instructions for the reception of the especially honored combatants in Spain by Hitler in the Marble Gallery of the *Reichskanzlei*.

German Fighter Ace Werner Mölders - An Illustrated Biography

Hitler at the welcoming of lined-up *Offiziere*, here from members of the *Kriegsmarine*.

From left to right: Göring, Hitler, Harlingshausen, Mölders, and unknown.

MOSAIKSAAL

TISCH 1: *der Führer*
General der Flieger Sperrle
General Don Antonioa Aranda
~~Vizeadmiral von Fischel~~ *General Ancipo de Leuns*
Oberst d. G. Warlimont
Oberstleutnant von Donat
Hauptmann Mölders
Leutnant Seiler
Oberfeldwebel Prestele
~~Feldwebel Bohasach~~

TISCH 2:
Generalfeldmarschall Göring
General der Flieger Volkmann
Admiral Boehm
General Don José Solchaga
Oberst Frhr. von Funk
Oberst Lindner
Oberstleutnant Plocher
Feldwebel Kroll
Feldwebel Flegel
Feldwebel Willuhn

TISCH 3:
Großadmiral Dr. h. c. Raeder
Konteradmiral Don Ramon Agacino y Armas
Admiral Carls
Kapitän zur See Cyliax
Korvettenkapitän Meyer
Korvettenkapitän Mark
Korvettenkapitän Erdmenger
Korvettenkapitän von Davidson
Kapitänleutnant Landheld
Kapitänleutnant Grosse

TISCH 4:
Generaloberst von Brauchitsch
General Don Carlos Martinez Campos
Oberst Schubert
Oberstleutnant Mehnert
Major Merhard von Bernegg
Hauptmann Hertzer
Oberleutnant Klinkicht
Oberfeldwebel Buchholz
Feldwebel Kahlert
Unteroffizier Köster

SPEISESAAL

TISCH 11:
General Don Juan Yague
Generalmajor Battisti
General d. Fl. z. D. Wilberg
Generalmajor Frhr. von Richthofen
Konteradmiral Fanger
Oberst Ritter von Thoma
Oberstleutnant von Donat
Major Neudörffer
Major Harlinghausen
Major Wolff
Hauptmann Schellmann
Hauptmann Harder
Oberleutnant Graf Hoyos
Oberleutnant Bertram
Oberleutnant Oesau
Oberleutnant Enzlen
Oberfeldwebel Hofmann
Oberfeldwebel West
Feldwebel Huth

TISCH 12:
Generalfeldmarschall Göring
General Don Rafael Garcia Valino
Oberst Prinz Alfonso de Orleans y Borbon
General d. Fl. Schweikhard
Oberst Kressmann
Oberstleutnant Seidemann
Oberstleutnant Holle
Major Handrick
Hauptmann Aldinger
Feldwebel Rochel
Feldwebel Brohammer
Unteroffizier Batz

TISCH 13:
Generaloberst Keitel
General Don Camilo Alonso Vega
Oberst Aurelia
Oberstleutnant Siber
Major Grabmann
Major Kuprian
Major Wäntig
Hauptmann Schröter
Hauptmann Knüppel
Oberfeldwebel Fischer
Oberfeldwebel Franz
Unteroffizier Beilharz

The seating arrangements for the mosaic hall and the dining hall on the occasion of the reception and the following meal in the *Reichskanzlei*.

German Fighter Ace Werner Mölders - An Illustrated Biography

After he brought his experiences in Spain to paper, and with that decisively contributed to the new fighter pilot rules it was time for a vacation. Mölders during breakfast in a pension in Garmisch-Partenkirchen.

Back in Wiesbaden as *Staffelkapitän* of the 1. *Staffel* of I./*Jagdgruppe* 133 (later I./JG 53 "Pik As"). To the far left Friedrich-Karl Müller (crashed fatally as bearer of the *Eichenlaub* in May 1944), and to the right of Mölders Hans von Hahn (*Ritterkreuz* on July 9, 1941).

The three successful fighter pilots of the "Legion Condor" (from the left): Schellmann, 12 air victories; Harro Harder, 11 air victories; and Mölders, 14 air victories. Unfortunately, all three died in World War II.

93

Missions on the Western Front

At the outbreak of war, Mölders was in the west with his *Staffel*, in order to protect the *Reich's* borders. Above: Mölders, unknown, and *Hptm*. Friedrich Beckh, who became his successor as *Geschwaderkommodore* in 1941, and has been missing since June 21, 1942.

German Fighter Ace Werner Mölders - An Illustrated Biography

The *Jagdstaffeln* of JG 53 "Pik As" remained in a so-called "*Sitzkrieg*" until May 1940. They were constantly on missions for border surveillance in their given section of Eifel-Saarbrücken.

German Fighter Ace Werner Mölders - An Illustrated Biography

The French fighter plane, Hawk 75 C-1, Curtiss "Mohawk." Mölders reported this as the first plane he shot down.

On September 20, 1939, Mölders won his 1st air victory, and received the *Eisernes Kreuz Zweiter Klasse* for it; afterwards the EK conferment in front of his Bf 109-E with the symbol of the JG 53 on the engine cowl.

The mention of the first downing of a plane in his flight book. However, the entry was mistakenly recorded under 21 September, while his first downing verifiably took place on 20 September! (See the "Air Combat Report" following the list of downings!)

This picture was taken during a short Christmas vacation in 1939 in Brandenburg/Havel.

By Klüsserath a.d. Mosel Mölders achieved his second shootdown on October 30, 1939, Bristol "Blenheim" bomber (in the above picture a unit of these English planes). Below: Members of the *Reichsarbeitsdienst* during an inspection of the remains of the plane.

German Fighter Ace Werner Mölders - An Illustrated Biography

A discussion of the mission with his pilots and a visit with the "men in black," whom he especially treasured. Top picture: fourth from the left *Uffz.* Neuhoff (Ritterkreuz on June 26, 1942, after 40 air victories), who received the certificate illustrated in the lower left.

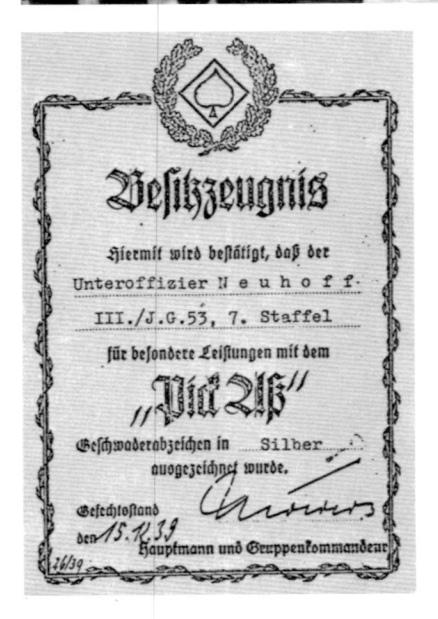

Mölders reports the readiness for duty of *Gruppe* III./JG 53 "Pik As" to his *Kommodore*.

German Fighter Ace Werner Mölders - An Illustrated Biography

On the Wiesbaden-Erbenheim air base shortly before the military campaign against France: the pilots of the *Gruppe* (top *Offiziere* and *Unteroffiziere*, below only the *Offiziere*). Many of these fighter pilots became "*Asse*" of their weapon. Mölders carries here already with the EK I that was awarded to him after his 7th air victory on April 2, 1940.

German Fighter Ace Werner Mölders - An Illustrated Biography

The conferment of decorations by Mölders to successful pilots (in the picture Ltn. Altendorf) and well-deserving members of the ground crew. Right: his *Adjutant*, Ltn. Radlick (fallen on October 2, 1940).

Humorous pictures and sayings on the *Gruppe*'s *Werkstattzug*, made by the loyal "men in black," the technicians of the weapons and planes, that posed here for the photographer for a souvenir photograph.

German Fighter Ace Werner Mölders - An Illustrated Biography

On May 10, 1940, the military campaign against France began. III./JG 53 carried out missions and achieved their first successes under the leadership of Mölders. Mölders in his Bf 109-E shortly before takeoff.

Description of air combat after returning from a mission. Mölders and his comrades listen eagerly to the successful pilot's report.

A British Hurricane unit above France. Mölders also defeated numerous planes of this type in hard air combat.

German Fighter Ace Werner Mölders - An Illustrated Biography

Mölders in the "*Spanien-Kluft.*"

On May 21, 1940, Mölders shot down three French Morane-*Jäger*. After landing he reported to the *Geschwaderkommodore* (October 1939-June 1940), *Gen.* Major Hans Klein (right). (Bearer of the "pour le mérite" medal for 22 air victories in World War I.)

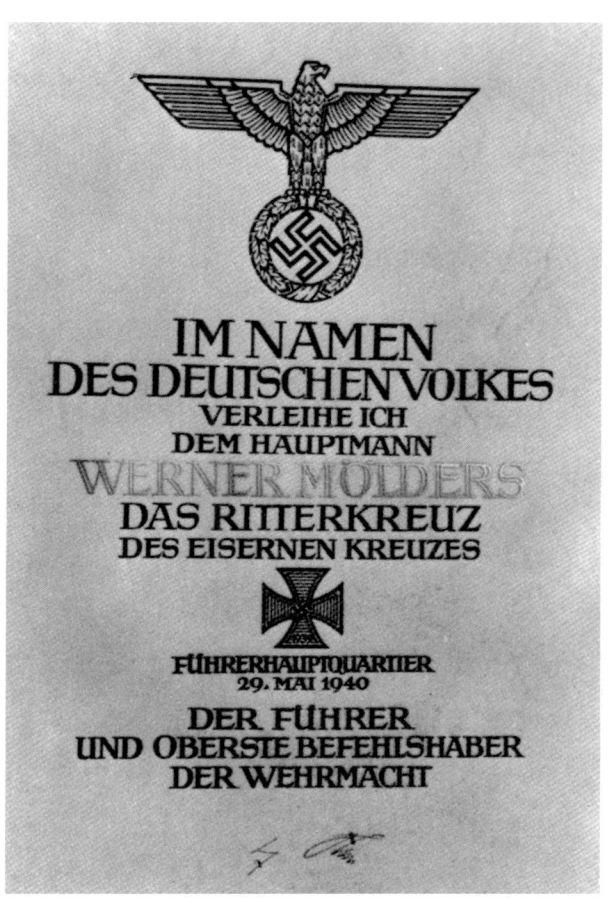

German Fighter Ace Werner Mölders - An Illustrated Biography

After his 20th air victory on May 27, 1940, in the region southwest of Amiens, he became the first fighter pilot awarded with the *Ritterkreuz des Eisernen Kreuzes* on May 29, 1940.

Two of many press releases of the 20th Air victory report:

> Die Verluste des Gegners in der Luft betrugen am 28. 5. insgesamt 24 Flugzeuge, davon wurden im Luftkampf 16, durch Flak acht abgeschossen. Drei deutsche Flugzeuge werden vermißt. Hauptmann Mölders errang seinen 20. Luftsieg.

> **Hauptmann Mölders**
> hat sich durch persönliche Tapferkeit vor dem Feinde sowie kühnen und besonders erfolgreichen Einsatz seiner Jagdgruppe ausgezeichnet. Hauptmann Mölders hatte bereits bis zum 29. Mai 20 Luftsiege errungen.

There are no pictures available of the conferment of the *Ritterkreuz* that took place in Göring's headquarters. The pictures show Mölders after his return to the *Gruppe*, newly decorated with the "*Dödel*," as the *Ritterkreuz* was called in the *Luftwaffe*. On the rudder assembly of his plane the lines marking shootdowns No. 19 and 20 for his achieved air victories on 27 May were missing.

The joy of his "heap" was naturally enormous, and the proud event was appropriately celebrated.

The official photograph of "*Vati*" Mölders, the first bearer of the *Ritterkreuz* in the German *Jagdwaffe*.

German Fighter Ace Werner Mölders - An Illustrated Biography

On May 31, 1940, Mölders was successful in shooting down a French plane LeO 45 in the Abbéville-Amiens region. Soldiers of the *Heer* who were in close proximity to the crash site took a few pictures and sent them to Mölders.

View of the JG 53 air field during the military campaign against France. In the background an A 34 "Hindenburg" used as an air ambulance lands.

Entered in Mölders' flight book under June 5, 1940, are the last two air victories that he achieved in the western military campaign.

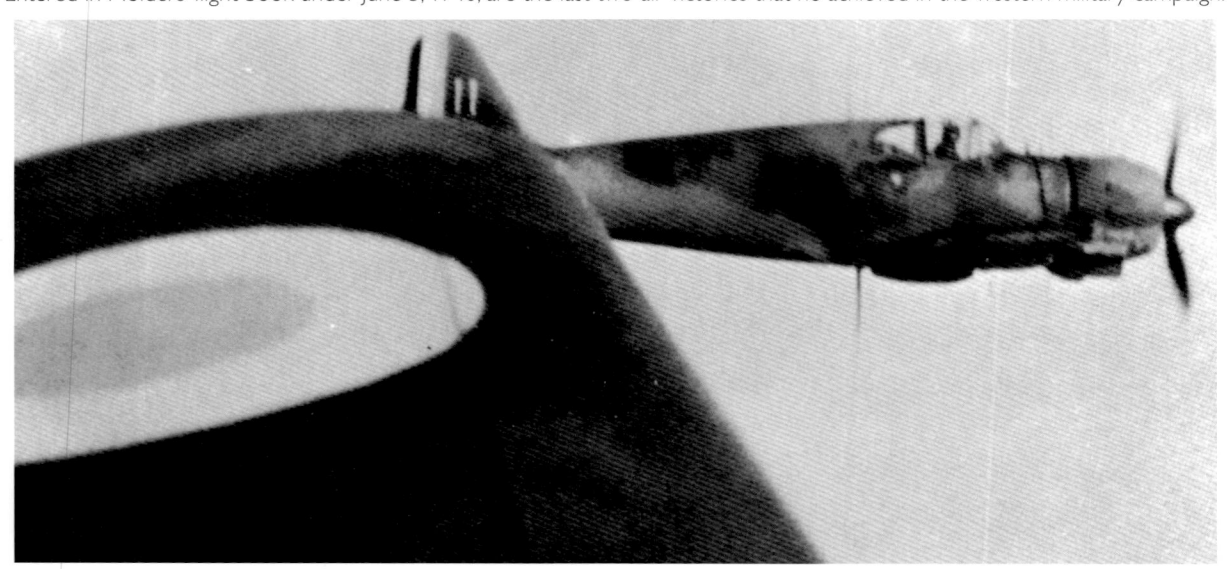

They were two French planes, a Bloch 152 fighter plane (24th shootdown)

... and a reconnaissance plane, Potez 63 (25th shootdown).

German Fighter Ace Werner Mölders - An Illustrated Biography

In this Bf 109 E-3, in which he achieved his two previously mentioned shootdowns, he was shot down in air combat by the French Ltn. Pommier-Layrargues.

One of the last photographs of Mölders, shortly before he was shot down and taken captive.

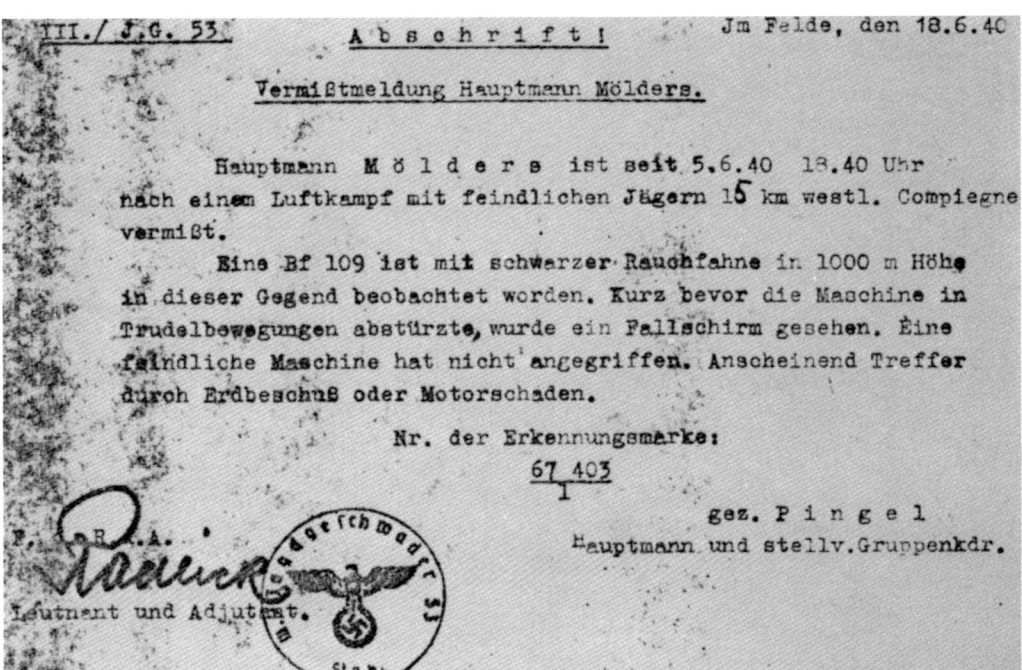

Mölders, who saved himself from the burning plane with a parachute, was then missing, as this report shows.

```
III./J.G. 53            Abschrift!            Jm Felde, den 18.6.40

             Vermißtmeldung Hauptmann Mölders.

        Hauptmann Mölders ist seit 5.6.40 18.40 Uhr
    nach einem Luftkampf mit feindlichen Jägern 15 km westl. Compiegne
    vermißt.
        Eine Bf 109 ist mit schwarzer Rauchfahne in 1000 m Höhe
    in dieser Gegend beobachtet worden. Kurz bevor die Maschine in
    Trudelbewegungen abstürzte, wurde ein Fallschirm gesehen. Eine
    feindliche Maschine hat nicht angegriffen. Anscheinend Treffer
    durch Erdbeschuß oder Motorschaden.

                        Nr. der Erkennungsmarke:
                              67 403
                              ─────
                                1
                                    gez. Pingel
                              Hauptmann und stellv.Gruppenkdr.
```

107

German Fighter Ace Werner Mölders - An Illustrated Biography

Above: Mölders' conqueror, Sous-Lieutenant Pommier-Layrargues (24 years old). During this air battle he brought down another Bf 109 and then crashed fatally with his Dewoitine 520 (No. 266). Fire on impact in the suburbs of Marissel by Beauvais. Right: The last entries in Mölders' flight book.

A Dewoitine 520, the best French fighter plane in World War II (910 PS-Motor, maximum speed 525 km, a 20 mm canon and 4 MG, and 1230 km range).

Missions on the Channel

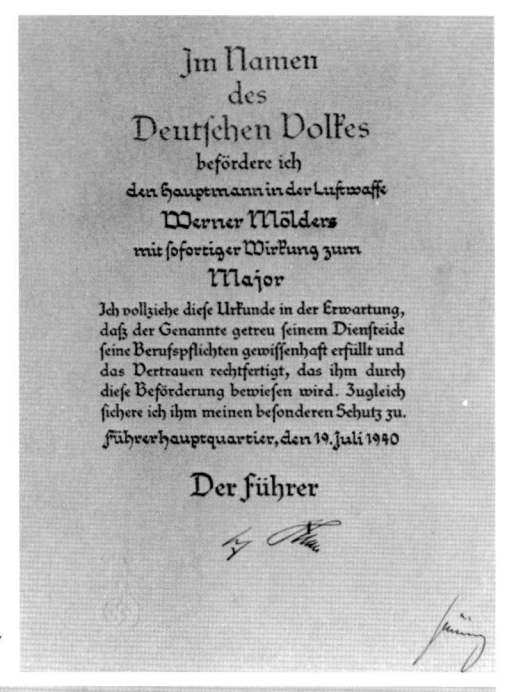

After his return from captivity (June 30, 1940), Mölders was soon promoted to *Major* on July 19, 1940.

On 20 July he was appointed to *Kommodore* of JG 51 and successor of Osterkamp. After his dismissal by the *Platzkommandant*, Major Schuster, he traveled with a Messerschmitt Bf 108 "Taifun" to the channel front.

After his arrival on the channel Mölders, among others, was well received by *Major* Adolf Galland, at this time *Kdr.* III./JG 26 "Schlageter" (*Geschwaderkommodore* as of August 22, 1940), where memories and experiences were exchanged.

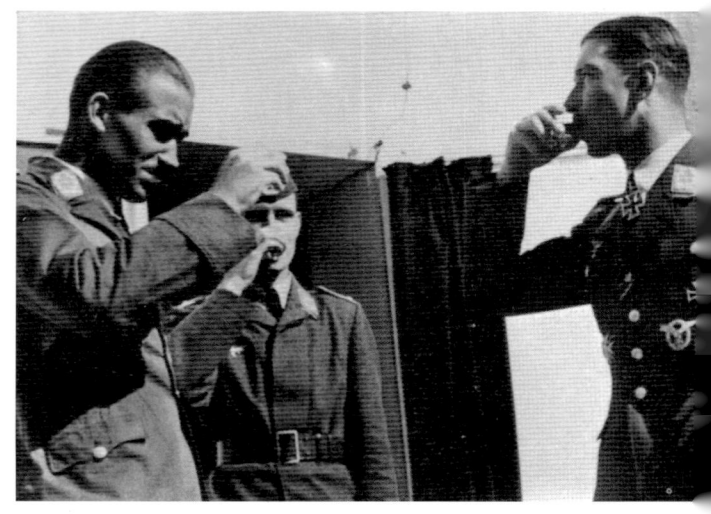

German Fighter Ace Werner Mölders - An Illustrated Biography

As of 28 July he was in missions on the channel and over England. Mölders shortly before takeoff, or rather, after his return from a mission in England. The traces of hard combat with English fighter pilots over the island were clearly shown in his face.

Sgt. John Burgess of the 222nd Squadron in his Spitfire ZD-P. On October 29, 1940, he shot down the Bf 109 of the bearer of the *Ritterkreuz*, Obltn. Otto Hintze, from 3./E-*Gruppe* 210. Hintze, who was captured, was an old comrade of Mölders from his time with the *Jagdgruppe* 134 in Werl.

German Fighter Ace Werner Mölders - An Illustrated Biography

A Do 17 Z bomber unit flies in the direction of England.

A war correspondent of the *Luftwaffe* drew the *Begleitjäger* Bf 109 and its protégé, the He 111 bomber, during their mission in England.

Right: During the "air battle for England," the *Reichsmarschall* observes the approach of his units and the air battles over the island from Cap Gris Nez. Next to Göring is *Gen. Feldmarschall* Kesselring, at that time head of *Luftflotte* 2.

German Fighter Ace Werner Mölders - An Illustrated Biography

A Rotte Bf 109 flies just above the channel waters to their target on the island.

Bf 109 E-4 near the steep English coast.

Bottom left: The Germans are coming! The alarm sounds on an English air field, and quickly they run to the planes.

Right: The vapor trails of German and English units are visible for miles around in the skies above England.

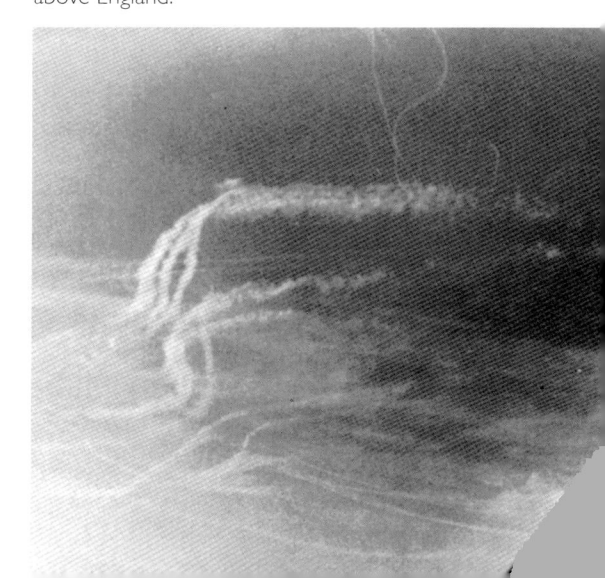

Successfully back from the mission in England! The *Messerschmittjäger* flies over the area in a swerving motion while Mölders and his "Katschmarek," *Obltn*. Claus, have already landed and taxi to a halt.

First words to his outstanding *Warte* and loyal men of the ground crew before he climbs out of the plane after landing.

German Fighter Ace Werner Mölders - An Illustrated Biography

Obltn. Claus also achieved a shootdown, and was cheerfully greeted by his men.

Surrounded by listeners, Mölders and Claus report on their mission over the island.

Hptm. Tietzen, *St.Kpt.* of 5./JG 51, did not return from his mission on August 18, 1940. He later washed ashore by Calais. He achieved 27 air victories, 7 of which were in Spain. Subsequently, he was honored with the *Ritterkreuz*.

German Fighter Ace Werner Mölders - An Illustrated Biography

Mölders' men knew and loved him as the beaming, laughing "*Vati*" Mölders.

Despite the difficult missions above the island and the channel, the pilots did not lose their sense of humor, as the drawing to the left on *Staffel*'s accommodations of JG 51 shows.

Right page: They protect their homeland! Hawker Hurricane of the 85th Squadron approaching the bomber and other units of the *Luftwaffe*.

The opponent is defeated again! It was the 32nd shootdown, achieved on August 31, 1940. Carefully, the Wart paints the 32nd line on the rudder assembly.

116

German Fighter Ace Werner Mölders - An Illustrated Biography

German Fighter Ace Werner Mölders - An Illustrated Biography

Mölders (to the far right) in excited conversation with his pilots shortly before takeoff for a mission. Everyone is already wearing their life jackets.

Back from a mission in England. Right—in a leather jacket—his brother Victor, St.Kpt. of JG 51 Jabo-Staffel (9 air victories as Zerstörer, participant in military campaigns in Poland, Norway, and France, as well as during Nachtjagd). He was shot down in a Jagdbomber mission on October 7, 1940, and was taken into English captivity.

German Fighter Ace Werner Mölders - An Illustrated Biography

As well as the *Kommodore*, all other comrades were cheerfully greeted after their return from the island, and they listened eagerly when they reported on their mission. Here an *Unteroffizier* of III./JG 51 whose name is, unfortunately, not known.

It was the same for JG 51 and other units of the *Jagdwaffe* who were sent to the channel. *Hptm.* Wick describes an air battle. As *Kommodore* of the JG 2 "Richthofen" he was shot down on November 28, 1940, after 56 air victories, and went missing after a parachute jump into the ocean south of Wight Island. Right *Obltn.* Leie, after 118 air victories, fell on March 1, 1945, in the east.

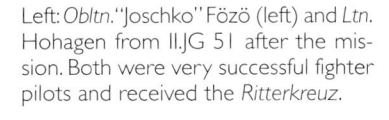

Left: *Obltn.* "Joschko" Fözö (left) and *Ltn.* Hohagen from II.JG 51 after the mission. Both were very successful fighter pilots and received the *Ritterkreuz*.

Right: EK I conferment by Mölders to successful pilots of his *Geschwader* (all *Unteroffizier* ranks). *Hptm.* Laumann (left) seems to check the authenticity of the medal beforehand.

German Fighter Ace Werner Mölders - An Illustrated Biography

September 1940: *Gen.Oberst* Udet, the greatest fighter pilot of World War I and excellent stunt pilot, visits his young fighter pilot comrades on the channel, who now are considered "*Asse*" of their weapon. From left to right: Balthasar, Oesau, Galland, Udet, Mölders, and Pingel. Only Galland and Pingel, who were taken into English captivity, survived the war!

For all of England's outstanding fighter pilots may the following opponent "*Asse*" be acknowledged. James Edgar Johnson ("Johnny" J.), *Geschwaderkommodore*, with 38 air victories, the greatest "As" of the RAF, and receiver of the highest awards.

German Fighter Ace Werner Mölders - An Illustrated Biography

Flight Lieutenant Mungo-Park, 27 air victories, fallen on July 1, 1941.

Bottom left: Wing Commander in Hawkinge (*Geschwaderkommodore*), Michael Nicolson Crossley, 22 air victories.

Right: Group Captain Adolphus Gysbert Malan ("Sailer" Malan). Together with the fallen Wing Commander Finucane (July 14, 1942), with 32 air victories he stands at second place out of the most successful fighter pilots of the RAF of World War II.

German Fighter Ace Werner Mölders - An Illustrated Biography

On September 20, 1940, Mölders shot down two Spitfires of the 92nd Sqn. by Dungeness, from which he became the first German fighter pilot to reach his 40th air victory in the war. For his efforts and the outstanding successes of his *Geschwader* he was the second soldier of the *Wehrmacht* (after General Dietl) to receive the *Eichenlaub zum Ritterkreuz des Eisernen Kreuzes* on September 21, 1940. The presentation of the award by Hitler followed on Sunday, September 23, 1940, in the new *Reichskanzlei* in Berlin.

dnb. Berlin, 23. Sept. Der Führer und Oberste Befehlshaber der Wehrmacht hat dem Major Mölders anläßlich seines 40. Luftsieges das Eichenlaub zum Ritterkreuz des Eisernen Kreuzes verliehen. Der Führer hat an Major Mölders das folgende Telegramm gerichtet: „In dankbarer Würdigung Ihres heldenhaften Einsatzes im Kampf für die Zukunft unseres Volkes verleihe ich Ihnen zu Ihrem 40. Luftsieg als zweitem Offizier der deutschen Wehrmacht das Eichenlaub zum Ritterkreuz des Eisernen Kreuzes."

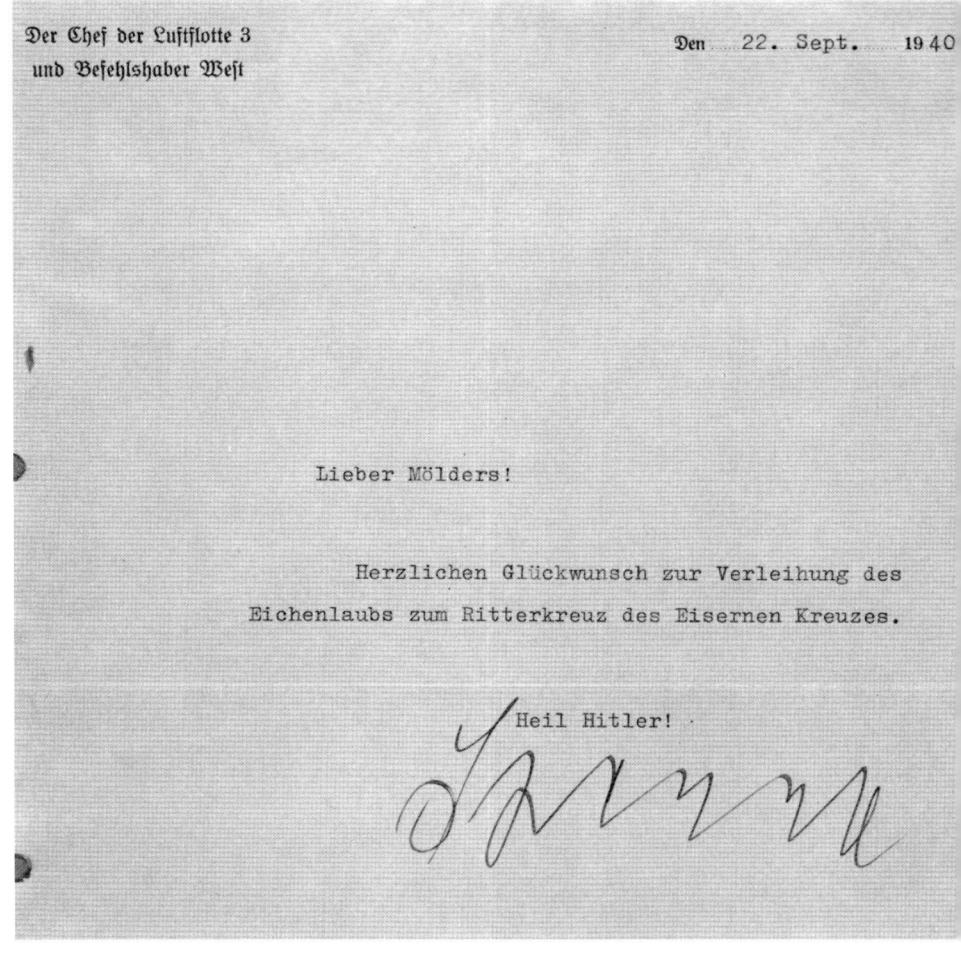

Der Chef der Luftflotte 3 und Befehlshaber West

Den 22. Sept. 1940

Lieber Mölders!

Herzlichen Glückwunsch zur Verleihung des Eichenlaubs zum Ritterkreuz des Eisernen Kreuzes.

Heil Hitler!

One of the many press reports, as well as a letter and telegram from the countless number of people that Mölders received for this event.

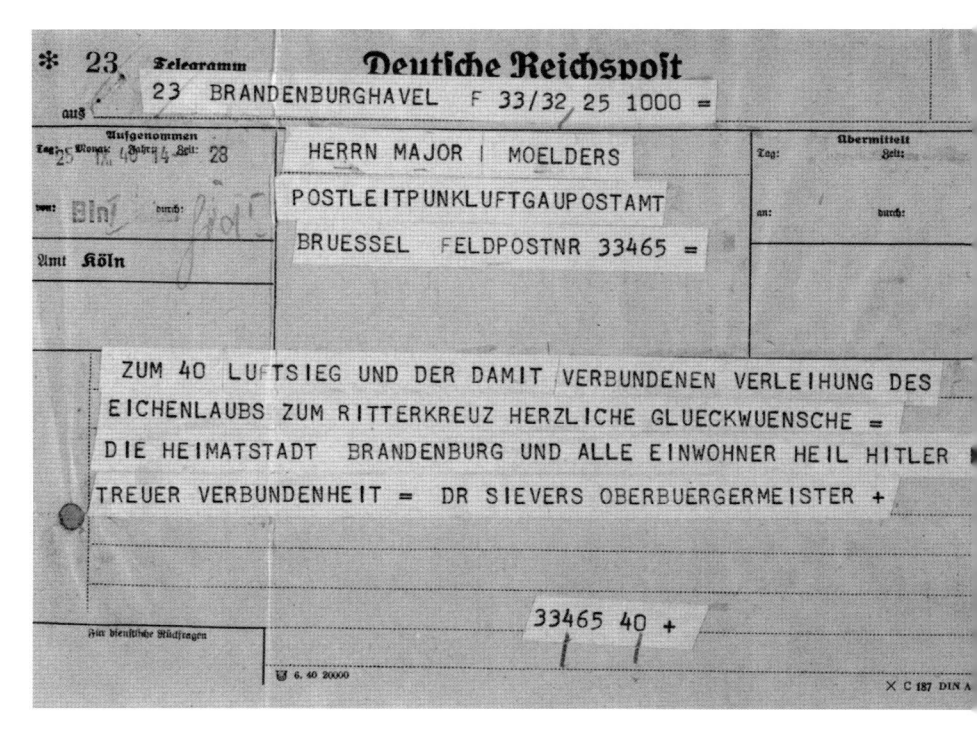

Telegramm — Deutsche Reichspost

23 BRANDENBURGHAVEL F 33/32 25 1000 =

HERRN MAJOR MOELDERS
POSTLEITPUNKT LUFTGAUPOSTAMT
BRUESSEL FELDPOSTNR 33465 =

ZUM 40 LUFTSIEG UND DER DAMIT VERBUNDENEN VERLEIHUNG DES EICHENLAUBS ZUM RITTERKREUZ HERZLICHE GLUECKWUENSCHE = DIE HEIMATSTADT BRANDENBURG UND ALLE EINWOHNER HEIL HITLER TREUER VERBUNDENHEIT = DR SIEVERS OBERBUERGERMEISTER +

33465 40 +

German Fighter Ace Werner Mölders - An Illustrated Biography

After the conferment of the *Eichenlaub* Mölders received an invitation from Göring, who was also *Reichsjägermeister*, to hunt. Mölders stayed in Göring's hunting lodge for a few days in the Rominter Heide. In the pictures above, *Gen. Feldmarschall* Milch in civilian clothes. In the picture below, left of Göring, is *Gen.Oberst* Ernst Udet.

German Fighter Ace Werner Mölders - An Illustrated Biography

The "*Reichsjägermeister*" and his escorts in front of the hunting lodge. Top picture: Mölders, left, with a shotgun strapped on. Bottom: Second from the right, with binoculars.

German Fighter Ace Werner Mölders - An Illustrated Biography

Mölders was already with his *Geschwader* on the channel again at the end of September. A welcoming by *Ltn.* Meyer, who also just returned from a successful mission over England. Later he was shot down and taken into English captivity.

With serious concentration, Mölders listens to the report of what occurred in his absence of the *Stabstaffel*. Left in the picture is the rudder assembly of his Bf 109 with lines marking shootdowns.

The *Reichsmarschall* visits the *Geschwader* on the channel. From left to right: *Ltn.* "Bubi" Kunze, *Hptm.* Mufl, Göring, *Gen.Oberst* Loerzer, and Mölders.

German Fighter Ace Werner Mölders - An Illustrated Biography

Mölders and *Hptm*. Oesau, at this time *Staffelkapitän* of 7./JG 51 Oesau, who was in Mölders' *Staffel* in Spain, fell in an air combat with a Lighting above the Eifel on May 11, 1944, as *Oberst* and *Kommodore* of the JG 1 after 125 air victories.

In a serious discussion with Günther ("Franz") Lützow, one of the most distinguished personalities of the German fighter pilots. Unfortunately, he also did not survive the war. As *Oberst* and member of JV 44—Galland—he has been missing since April 24, 1945, with his Me 262 in the Donauwörth region.

Meeting of *Kommodores* at the *Reichmarschall*'s. From left to right: *Major* Bloedorn, KG 4; *Major* Storp, KG 76; *Oberstltn*. Galland, JG 26; *Reichsmarschall* Göring; *Oberstltn*. Mölders, JG 51; and *Major* Wick, JG 2.

Josef ("Pips") Priller was regarded as one of the many outstanding German fighter pilots who, because of Mölders, became experts of their weapon. His *Staffel*, 6./JG 51, drew a congratulatory poster for him bottom left on the occasion of his 20th air victory and the conferment of the *Ritterkreuz* on October 19, 1940. Priller achieved a total of 101 air victories in the west, 11 of which were Viermots. He was one of the few who achieved over 100 shootdowns in the west.

Above: His men not only respected Mölders as a man and officer. Abroad, especially in Spain, he was held in great esteem, which many letters he received from there prove. One of these is presented here.

German Fighter Ace Werner Mölders - An Illustrated Biography

The "men in black" during servicing of his plane. He had an especially good relationship with and a good ear for the problems of his "black boys," as he would say. It is no wonder that they admired him as "their *Vati*," just as the men of his flying personnel did.

51 lines marking shootdowns on his rudder assembly! He gained his last three air victories on October 22, 1940, NW Maidstone (No. 49-51). The shootdowns in Spain were not listed because the pilots were prohibited to paint them on.

Mölders zum Oberstleutnant befördert

Berlin, 26. Oktober.

Der Führer hat auf Vorschlag des Oberbefehlshabers der Luftwaffe, Reichsmarschall Göring, den Geschwaderkommodore Major Mölders anläßlich seines 50. Luftsieges wegen besonderer Tapferkeit und seiner großen Verdienste um die Schlagkraft der deutschen Jagdfliegerei bevorzugt zum Oberstleutnant befördert. Oberstleutnant Mölders erzielte am Freitag seinen 52. und 53. Luftsieg.

Der Führer hat mit Wirkung vom 10. Oktober den Oberstleutnant i. G. Hanns Seidemann zum Oberst befördert.

After his 50th air victory Mölders was promoted to *Oberstleutnant*. Gen.Major Osterkamp ("Uncle Theo"), head of the fighter pilots on the channel, congratulates him and reports on the last mission. Among the circle of listeners and well wishers, Walter Oesau (picture center), *Kommandeur* of III./JG 51.

Italian *Fliegeroffiziere* who visit German fighter pilots on the channel arrive a short time later to congratulate him. Mölders in a conversation with one of his Italian comrades.

German Fighter Ace Werner Mölders - An Illustrated Biography

At the end of October 1940 *Gen.Feldmarschall* Kesselring visited JG 51. In the background his *Reiseflug*, a Siebel Si 104 with a raised *Kommando* flag of a *Luftflotte* leader on the plane.

Kesselring in a conversation with one of the pilots. To the left of Mölders, *General d.Flg.* Wenninger, bearer of the "Pour le Mérite" of World War I, that he received on March 10, 1918, as a successful *U-Boot-Kommandant*. (He switched over to the *Luftwaffe* before the outbreak of the war in 1939.)

In great spirits, Kesselring, Mölders, and the *Geschwaderstab* pose for the photographer for a souvenir photograph. Second from the left: *Feldwebel* Fleig, Mölders' *Rottenflieger*.

131

German Fighter Ace Werner Mölders - An Illustrated Biography

Once again it went well! Mölders after his return from a mission in England with shots in the cabin that missed his head by an inch.

One victory followed the other. The rudder assembly on October 29, 1940, after 54 shootdowns.

A delightful conversation with his "Katschmarek," *Feldwebel* Erwin Fleig. Awarded with the *Ritterkreuz* and promoted to *Leutnant*, Fleig, after 66 air victories, had to bale out after his plane was hit on May 29, 1942, by Szokoloje (Russia), and was taken into captivity for many years after his parachute jump.

German Fighter Ace Werner Mölders - An Illustrated Biography

Mölders in his combat post on the field telephone.

On November 11, 1940, Mölders lost his best friend and comrade, *Oblt.* "Schorse" Claus, at this time *St.Kpt.* of 1./JG 51 (Claus next to Mölders in the picture above right). After 27 air victories he did not return from an escort assignment after air combat with Spitfires above the mouth of the Thames and has been missing since.

Weeks before his brother Victor was taken prisoner. The photograph to the left shows a very serious *Kommodore*.

133

German Fighter Ace Werner Mölders - An Illustrated Biography

For compensation between the difficult missions over England and the channel that involved heavy loses, now and then the fighter pilots went hunting. As can be seen here, with much success. In the top picture to the left behind Mölders *Obltn*. Leppla, to the right of Mölders "Hannes" Trautloft and Rolf Pingel.

German Fighter Ace Werner Mölders - An Illustrated Biography

Two more snapshots of the hunting excursion in which *Oberstltn*. Galland (bottom picture in the center) took part. In the top picture next to Mölders is Rolf Pingel, and to the right "Hannes" Trautloft, who can also be seen in the bottom picture. In this picture to the far right, Frhr. von Maltzahn and outside left, Herbert Wehnelt.

```
R a m c k e , Oberst.           Zeithain, den 4. Dez. 1940.


        An den
        Geschwaderkommodore
        Herrn Oberstleutnant  M ö l d e r s ,
        Ritter des Ritterkreuzes zum Eisernen Kreuz
                        mit Eichenlaub.


        Lieber Herr Mölders!

              Unlängst traf ich bei einem Dienst in der Fallschirm-
        truppe, der ich seit Juli d. J. angehöre, den Major und Gruppen-
        kommandeur Erdmann, der sich z. Zt. in Tutow befindet und wir
        erinnerten uns mit Freude und stolzer Genugtuung des einstigen
        Fahnenjunkers in der 11. Komp. des I.R. 2 in Lötzen, deren alter
        Chef ich damals war.
              Wenn Sie, lieber Herr Mölders, auch nur eine kurze Gast-
        rolle bei meiner Komp. gegeben haben, so erinnere ich mich ge-
        nauestens des ganz besonders frischen und strammen Fahnenjunkers
        Mölders und ich bin stolz darauf, daß Sie als nunmehr weit über
        Deutschland hinaus bekannter Jagdflieger, in meiner Komp. als
        Fahnenjunker Dienst getan haben.
              Ich nehme daher, wenn auch verspätet, in aufrichtiger
        Freude Gelegenheit, Ihnen zu Ihrer ruhmvollen Laufbahn in der
        Luftwaffe und zu Ihren glänzenden Erfolgen  im Kampfe in Spanien,
        Frankreich und über England meine allerherzlichsten Glückwünsche
        darzubringen. Möge ein gütiger Himmel Ihnen weiterhin so große
        Kampferfolge und damit Ruhm und Ehre beschieden sein lassen und
        vor allem dahin gnädig sein, daß Sie nach dem Endkampf
        und Sieg über den Engländer, gesund in Ihre Heimat zurück-
        kehren können.
                In kameradschaftlicher Verbundenheit bin ich
```

Oberst Ramcke, the writer of this letter, was *General* of the *Fallschirmtruppe* in 1944, and was awarded, like Mölders, with the "*Brillant.*" There is another parallel between him and Mölders. Like him, Ramcke was greatly honored and respected by his men, and called "*Vater* Ramcke!"

For the Yule festival (winter solstice) there was a great festival for the fighter pilots on the channel on 21 and 22 December 1940, and Mölders received a beautiful sword and the certificate from the "*Kanalkampfführer*" (*Kanalkafü*) "Uncle Theo" Osterkamp presented here.

German Fighter Ace Werner Mölders - An Illustrated Biography

Above: A ski trip in Zürs am Arlberg: cheerful, beaming faces. the horse-drawn sled goes from the hotel to the runway. Next to Mölders in the sled, *Hptm.* Balfanz. Below: In this magnificent landscape, on the white Arlberg, the men of JG 51 recuperate from the strains of difficult missions on the channel.

Mölders was an enthusiastic skier, and was always on his skies. He also had more than his fair share of fun, as one can see, he lets himself be put on by the smart "ski bunnies."

During this trip Mölders was invited a number of times to hunt in the snow covered mountain forests by Lech am Arlberg. These pictures were taken during the "Pirschgänge" on skis.

German Fighter Ace Werner Mölders - An Illustrated Biography

In front of the "Zürserhof," putting on their skis. *Mutti* Mölders, who came for a visit, gives good pieces of advice on the way.

Bottom left: Cheerfully "talking shop" with his *Rottenflieger* Erwin Fleig (right), and with the Olympic champion (1936) and world champion (1937-39), Christl Cranz-Borchers. In the center Karlfried Nordmann, later bearer of the *Eichenlaub* and *Kommodore* of JG 51 "Mölders."

German Fighter Ace Werner Mölders - An Illustrated Biography

On January 9, 1941, the community of Zürs organized a cheerful, festive evening for Mölders and his comrades in the Hotel "Alpenrose."

Saying goodbye. Mölders leaves Zürs and returns to the channel front. From left to right: "*Fürst*" Wilcke, Mölders, unknown, and "Tutti" Müller.

A short stopover at Wiesbaden-Erbenheim on the way to the front. The *Geschwader*'s parked Bf 109-Es and a Go 145 on the field.

Mölders' handwritten dedication in the *Offiziersheim*'s guest book on the Wiesbaden-Erbenheim air field from January 22, 1941:"Eyes open, approach, and calmly shoot! Werner Mölders."

February 1941 on the channel; the successful pilots of 12./JG 51 (from left to right): *Ofw*. Heinrich Hoffmann, *Ofw*. Fleischhacker, *Obltn*. Nordmann, *Ltn*. Beise, *Feldw*. "Peppi" Jennewein, and *Ltn*. Wiest. Only Beise and Nordmann survived the war.

German Fighter Ace Werner Mölders - An Illustrated Biography

Riding and shooting were popular recreational sports in between difficult missions. The *Kommodore* on the "1 PS-Hafermotor" on the channel coast (one of the rarest shots of Mölders!), and during KK-shooting with *Obltn*. Stengel, *Hptm*. Fözö, and *Ltn*. Rübell (second from right).

Mölders visits wounded comrades in the *Luftwaffe*'s military hospital, and is photographed with Sister Frida (left) and Sister Martha for a souvenir album.

Messerschmitt Bf 109 E-4 before take-off.

Mölders climbs into his plane, and after the usual takeoff preparations with his *Warte*...

...he takes off with absolute concentration for a mission over the channel to England.

Shootdown! The 60th air victory has been achieved! This on-board camera captures this moment, and shows the burning Spitfire that Mölders shot down SO Dungeness at an altitude of 6500 m on February 26, 1941.

A happy and victorious return. The beaming *Kommodore* after the 60th air victory.

German Fighter Ace Werner Mölders - An Illustrated Biography

On "*Helden-Gedenktag*," March 12, 1941, Mölders visited a military cemetery, and placed a wreath on the grave of a fallen comrade.

Reichsmarschall Göring in a conversation with Mölders on the occasion of a *Kommandeur* meeting in Paris. Göring looks at the "*Spanien-Kreuz in Gold mit Schwertern und Brillanten*" that was awarded a total of 28 times to members of the "Legion Condor."

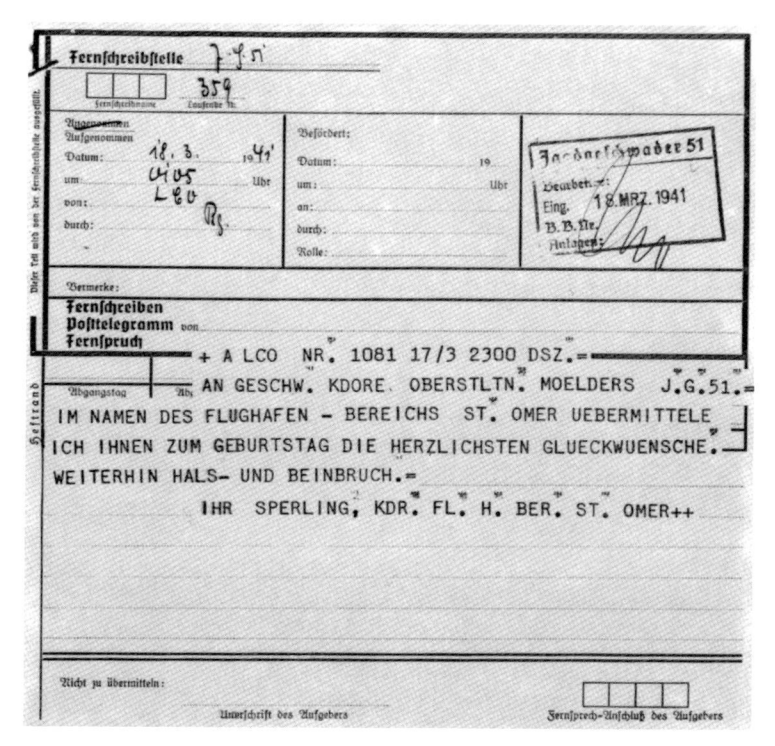

German Fighter Ace Werner Mölders - An Illustrated Biography

On his 28th birthday Mölders received a flood of letters and telegrams. No one knew then that it would be his last.

Since January 1941 *Obltn.* Hartmann Grasser flew with Mölders in the *Geschwadersstabsschwarm*. Like many others who flew with "*Vati*" Mölders, he was one of the best of the German *Jagdwaffe* (103 air victories on 700 missions, bearer of the *Eichenlaub zum Ritterkreuz des Eisernen Kreuzes*).

Return from a mission over the island. His true "black boys" are always the first ones by him.

The *Kommodore's Wartungsgruppe* during the service of his plane. He always spoke of them and their work with high praise.

German Fighter Ace Werner Mölders - An Illustrated Biography

April 15, 1941: The "*Jafü*" 2, *Gen.Major* "Uncle Theo" Osterkamp's birthday. His *Kommodore* came to congratulate him and to report. From left to right: Lützow, JG 3; Galland, JG 26; Osterkamp; von Maltzahn, JG 53; Mölders, JG 51; and unknown.

The "experts" used this opportunity to exchange vivid accounts of air combat, as these pictures of Mölders and Galland show.

German Fighter Ace Werner Mölders - An Illustrated Biography

Mölders' official car was a "Wanderer"-Kabriolett that he enjoyed very much and often drove, like here in this picture. His driver sits in the back seat.

A prominent visit! *Oberstltn.* Galland, *Kommodore* of JG 26 "Schlageter," just landed. The *Geschwader*'s coat of arms and the "Micky Mouse," Galland's personal distinguishing mark on his Bf 109, are easily recognized. His beloved cigars, which could also be called "his distinguishing mark," are also present.

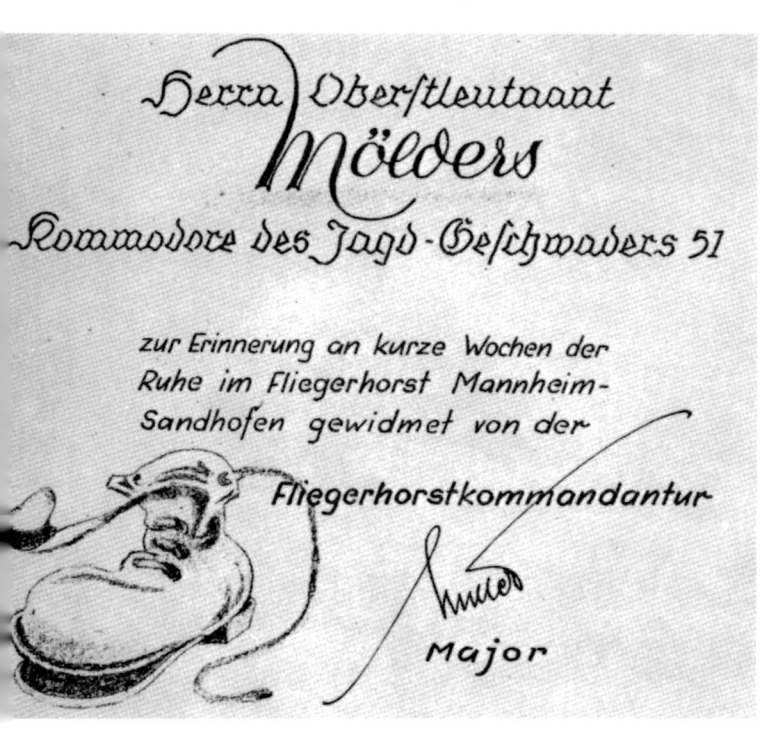

German Fighter Ace Werner Mölders - An Illustrated Biography

After a section of JG 51 was pulled from the channel, Mölders received a transfer command that ordered him and a large part of the *Geschwader* back home —for the time being, to Mannheim-Sandhofen.

May 19, 1941: Mölders just landed on the Mannheim-Sandhofen air base.

The festively decorated entrance to the air base.

In conversation with a few *Offiziere*. Second from the right, Wilhelm Moritz, in 1944 *Kommandeur* of the famous "*Sturmgruppe*", IV./JG 3 "Udet," especially successful in the "*Reichsverteidigung*."

On the occasion of JG 51's return home there was a large line-up to welcome them to the Mannheim-Sandhofen air base with a speech from *Oberstltn*. Mölders.

German Fighter Ace Werner Mölders - An Illustrated Biography

This picture was taken in the air base's *Offiziersheim*.

Wherever Mölders turned up, he suddenly became the focus of public attention. Here, after a visit of the *Luftgaukommandos* XII/XIII in Wiesbaden.

German Fighter Ace Werner Mölders - An Illustrated Biography

Ordered to Düsseldorf for a short time, he and his wife visited a football championship game of his favorite team, FC Schalke 04, in the Düsseldorf stadium.

His wife, Luise, took this fine picture, probably one of the best and most well known picture of Mölders in the office of the Düsseldorfer LW-Dienststelle.

German Fighter Ace Werner Mölders - An Illustrated Biography

On June 10, 1941, this nice time came to an end, and Mölders flew to his *Geschwader* in East Prussia. The *Kommodore*'s new Bf 109 F-2 on the Düsseldorf airport. The *Geschwader*'s coat of arms, the *Geschwader-Kommodore*'s distinguishing feature, and the lines marking shootdowns for 68 won air victories in the west are easily recognized.

Preparations for takeoff. Left in the picture (in the *Heer* uniform) is Fritz von Forell, author of Mölders' biography, "Mölders and his Men"; next to him is Mölders' nephew, Hartmut von Forell.

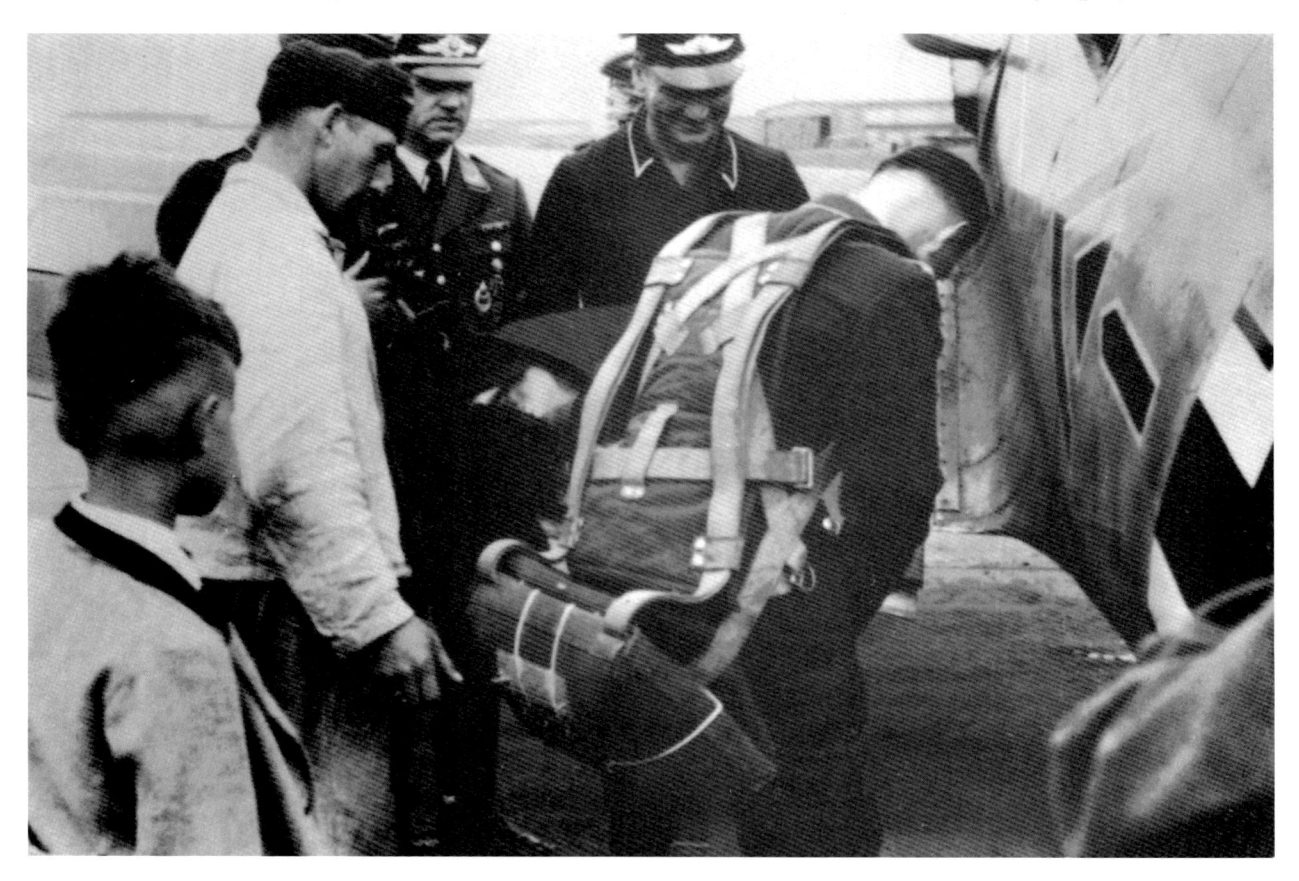

German Fighter Ace Werner Mölders - An Illustrated Biography

The last photographs before takeoff to the east, where there soon is a new, grave theater of war.

Missions on the Eastern Front

June 22, 1941: Mölders has just landed and reports, still sitting in the cockpit of his BF 109, on the 1st air victory (an I-153) that he achieved on the new front.

German Fighter Ace Werner Mölders - An Illustrated Biography

Mölders schoß den 72. Gegner ab

dnb. Berlin, 23. Juni.

Oberstleutnant Mölders schoß am 22. Juni seinen 72. Gegner in der Luft ab. Der Führer hat ihm aus diesem Anlaß das Eichenlaub mit Schwertern zum Ritterkreuz des Eisernen Kreuzes verliehen und ihm nachfolgendes Telegramm gesandt: „Zu Ihrem heute erfochtenen Luftsiege übermittle ich Ihnen meine besten Wünsche. In Ansehen Ihres immer bewährten Heldentums verleihe ich Ihnen als zweitem Offizier der deutschen Wehrmacht das Eichenlaub mit Schwertern zum Ritterkreuz des Eisernen Kreuzes. (gez.) Adolf Hitler

On the first day of the mission in Russia he shot down 3 more planes ("Martin-Bomber," SB-2), and for this received the *Eichenlaub mit Schwertern zum Ritterkreuz des Eisernen Kreuzes* as 2nd soldier of the *Wehrmacht* (after Galland).

After his return from the mission: exchange of experiences with Erich Hohagen (left) and "Schorsch" Eder (center). Both were, in the course of the war, highly decorated, and regarded as the most successful German fighter pilots.

German Fighter Ace Werner Mölders - An Illustrated Biography

A few photographs of the Russian planes that Mölders defeated during his 33 air victories in the east. Top left a "Rata," I-16. Right a I-153.

SB-2 bis, "Martin-Bomber."

PE-2

Top: A pilot's fate! The end of a brave opponent. A shot down I-153.

Center: This is what it looked like on the evening of June 22, 1941, after the first day of the mission on many Russian airfields! Destroyed Soviet planes wherever one looked.

On the morning of the second day of the campaign against Russia. From left to right: Grasser, Beckh (successor of Mölders as *Geschwader-Kommodore*), and to the far right, *Obltn*. Führing.

German Fighter Ace Werner Mölders - An Illustrated Biography

Will things go well? The *Kommodore* and a few comrades eagerly observe the landing of a pilot who is returning from a mission with a damaged plane.

Mölders during an inspection flight in the cockpit of a Siebel 104 "Halore."

Heartfelt congratulations for Hartmann Grasser for his new air victory.

German Fighter Ace Werner Mölders - An Illustrated Biography

His concern was for all soldiers! Here in conversation with three wounded *Landser* whom he temporarily took in before they came into the military hospital.

Congratulations for *Ofw.* Edmund Wagner from the 9. *Staffel*, who was soon regarded as the most successful pilot of JG 51 in Russia.

Mission discussion with bearer of the *Eichenlaub Hptm.* Joppien, *Kdr.* of the I. *Gruppe*. Only a few weeks later, on August 25, 1941, he crashed fatally by Brjansk in air combat. He had a total of 70 air victories, 42 of which were achieved in the west.

German Fighter Ace Werner Mölders - An Illustrated Biography

Visit of the *Geschwaderstab* JG 53 "Pik As" in the northern division in Krzewica. Second from the right, the *Kommodore* of JG 53, Frhr. von Maltzahn.

JG 51 was constantly on missions, and supported the *Panzer* and *Infanteriespitzen* wherever they went. At an arrangement of common operations with *Gen.Oberst* Guderian, at this time *Befehlshaber* of the *Panzergruppe* 2.

German Fighter Ace Werner Mölders - An Illustrated Biography

After victorious air combat in his Bf 109 F-2.

Mölders shot down 10 more opponents until July 4, 1941, and already had 82 shootdowns during the conferment of the "*Schwerter*," many of which were SB-2 bis "Martin-Bombers" (see picture).

German Fighter Ace Werner Mölders - An Illustrated Biography

The official press reports from the "*Völkischer Beobachter*" about the conferment of the high decoration by Hitler.

OBERSTLEUTNANT GALLAND  **OBERSTLEUTNANT MÖLDERS**

An der Luftschlacht des 21. Juni 1941 hat Oberstleutnant Galland durch hervorragende Führung seines Geschwaders und durch Abschuß von drei Gegnern besonderen Anteil gehabt. Oberstleutnant Mölders schoß am Tage darauf seinen 72. Gegner in der Luft ab. Der Führer hat aus Anlaß dieses heldenhaften Einsatzes beiden Offizieren das von ihm neugestiftete Eichenlaub mit Schwertern zum Ritterkreuz des Eisernen Kreuzes verliehen und ihnen ein entsprechendes Glückwunschtelegramm gesandt

The announcement of the conferment of the "*Schwerter*" to Galland and Mölders in the *Luftwaffen-Illustrierte* "*Der Adler.*"

On 3 July (the press report followed a day later!) Hitler received *Oberstltn.* Mölders in his headquarters, the "Wolfsschanze," and presented him the "*Schwerter.*" (The FHQu. was located a few kilometers east of Rastenburg/East Prussia, and was named "Wolfsschanze.")

165

German Fighter Ace Werner Mölders - An Illustrated Biography

Mölders shortly after his return from the *Führer's* headquarters, wearing the new decoration.

In a conversation with *Major* Friedrich Beckh, *Kdr.* of IV. *Gruppe*. Beckh took over the *Geschwader* a few weeks later.

In front of the "bookkeeping," the *Geschwader's* board of shootdowns.

German Fighter Ace Werner Mölders - An Illustrated Biography

The following three pilots of the *Geschwader* represent the memory of all fallen and missing JG 51 comrades:

Obltn. Heinrich Höfemeier, named "the thick one," *St.Kpt.* of 3./JG 51, 96 air victories, bearer of the *Ritterkreuz*, shot down on August 7, 1943, by flak near Karatschew.

Ofw. Heinrich Hoffmann, 12./JG 51, missing since October 3, 1941, 63 air victories, posthumously awarded with the *Eichenlaub zum Ritterkreuz*.

Ofw. Josef "Pepi" Jennewein, 2./JG 51, 86 air victories; crashed on July 26, 1943, on the other side of the front East of Orel after air combat and since missing. Awarded with the *Ritterkreuz* posthumously. (One of the best German skiers, two-time world champion in the Alpine Combination.)

German Fighter Ace Werner Mölders - An Illustrated Biography

Mölders in conversation with *Uffz.* "Toni" Hafner, the most successful pilot of JG 51. (After 204 air victories he fell in East Prussia as *Obltn.* and *St.Kpt.* of 10./JG 51 on October 17, 1944.) Picture center: "Joschko" Fözö.

"*Vati*" Mölders as a gentleman! Chocolates for both sweet DRK nurses of the casualty clearing station that set up its tents in close proximity to the combat post.

Fighter pilots must also sleep! A "little nap" under a shady tree in the midday heat of the hot Russian summer of 1941.

German Fighter Ace Werner Mölders - An Illustrated Biography

Victory after victory! Shootdown of a "Rata," I-16, that crashes like a burning torch, and ends with fire on impact.

July 15, 1941: Mölders has just landed. He was the first fighter pilot of the world to reach his 100th and 101st air victories (not including the shootdowns in Spain). Soldiers of all ranks and units near the field closely surround him.

With expressive gestures, Mölders illustrates the successful air battle to his listeners.

The congratulatory plaque from his men for his 100th air victory.

Des Führers Dank an Mölders

Der Führer und Oberste Befehlshaber der Wehrmacht hat folgendes Handschreiben an Oberstleutnant Mölders gerichtet:

Führerhauptquartier, 15. Juli.

Herrn Oberstleutnant Mölders, Kommodore.

Nehmen Sie zu Ihren heutigen fünf neuen Luftsiegen meine aufrichtigsten Glückwünsche entgegen. Sie haben mit diesen Erfolgen im großdeutschen Freiheitskampf 101 Gegner in der Luft abgeschossen und sind einschließlich Ihrer Erfolge im spanischen Bürgerkrieg 115 mal Sieger im Luftkampf gewesen.

In Würdigung Ihres immerwährenden heldenmütigen Einsatzes im Kampf um die Freiheit unseres Volkes und in Anerkennung Ihrer hohen Verdienste als Jagdflieger verleihe ich Ihnen als erstem Offizier der deutschen Wehrmacht die höchste deutsche Tapferkeitsauszeichnung, das

Foto: PK.-Jütte (Atlantic)

Eichenlaub mit Schwertern und Brillanten zum Ritterkreuz des Eisernen Kreuzes.

Mit meinem und des ganzen deutschen Volkes Dank verbinde ich die besten Wünsche für Ihre Zukunft.

gez. Ihr Adolf Hitler.

1200 Luftsiege des Jagdgeschwaders Mölders

* Stuttgart, 17. Juli.

Die Nachricht von der Verleihung des Eichenlaubs mit Schwertern und Brillanten zum Ritterkreuz des Eisernen Kreuzes an Oberstleutnant Mölders für 101 Abschüsse in diesem Krieg hat in der Bevölkerung ein verständlich freundliches Echo gefunden. Und nun überrascht uns die weitere Meldung, daß das Jagdgeschwader Mölders seit Beginn der Kampfhandlungen im Osten bis zum 12. Juli 500 Sowjetflugzeuge bei nur drei Verlusten abgeschossen hat. Das Geschwader hat damit am 12. Juli insgesamt den 1200. Luftsieg erkämpft.

Man fragt unwillkürlich, was sind das für Männer, die Deutschlands Jagdwaffe beim Gegner so gefürchtet und dadurch zu den beliebtesten Soldaten im deutschen Volk machen. Da erinnert man sich an das Bild in einer Illustrierten aus den Wintertagen; es zeigte sportgestählte und braungebrannte Männer bei einer Schneeballschlacht in den Bergen. Offiziere des Jagdgeschwaders Mölders mit ihrem Kommodore an der Spitze verbringen ein paar Erholungstage. Ein Blick in ihre Gesichter zeigt uns, was wir wissen wollen, um diese gewaltigen Erfolge zu verstehen; ein besonderer Typ von Männern, die ihre Siege nicht einem blinden Draufgängertum verdanken, sondern wohl überlegten Grundsätzen. Man weiß von Mölders, er sichert aufs sorgfältigste, dann pirscht er den Gegner an, überblickt die Situation und nähert sich endlich aus einer völlig geklärten Lage dem Gegner, so daß ihm als Ergebnis all dieser Ueberlegungen schließlich der Sieg sicher ist.

Kriegsberichter haben Mölders oft auszufragen versucht, aber es ist ihnen nicht immer gelungen. Eine große Bescheidenheit ist ein weiteres besonderes Kennzeichen des Fliegerhelden. Wenn Mölders allerdings einmal erzählt, dann spricht er nicht anders, als man dies von ihm erwartet. Er sagt zum Beispiel in der Schilderung eines Kampfes: „Ich haue ab, daß sich die Balken biegen."

Seine Schilderung über seinen 26. Abschuß beim Anblick einer Gruppe von sechs bis zehn Spitfires kleidet er in die Worte: „Ich bekomme einen Riesenschreck, weiß aber gleich, hier kann nur eins helfen, mitten durch den Haufen Engländer durchzustoßen." Durch Splitter ist Mölders übrigens gerade in diesem Kampf schon einmal verwundet worden. Im übrigen ist bekannt, daß er mit 14 bestätigten und mehreren unbestätigten Abschüssen der erfolgreichste Jagdflieger der Legion Condor in Spanien war, den neben dem Spanienkreuz in Gold mit Brillanten die Medalla de la Campana und die Medalla Militar schmücken. Mölders, heute 27 Jahre alt, ist in Gelsenkirchen geboren als Sohn eines Studienrates, der 1915 als Reserveoffizier fiel. Ursprünglich Infanterist, trat er 1935 bei der Neugründung in die Luftwaffe ein.

For this outstanding achievement Mölders receives as the first soldier the new and, until 1945, highest German medal for bravery, the *Eichenlaub mit Schwertern und Brillanten zum Ritterkreuz des Eisernen Kreuzes*. The conferment of the "*Brillant*" in the daily news.

> BERLIN, DEN
> Führerhauptquartier, den 15. Juli 1941.
>
> **ADOLF HITLER**
>
> Herrn
>
> Oberstleutnant M ö l d e r s ,
> Kommodore Jagdgeschwader 51
>
> Nehmen Sie zu Ihren heutigen 5 neuen Luftsiegen meine aufrichtigen Glückwünsche entgegen. Sie haben mit diesen Erfolgen im Grossdeutschen Freiheitskampf 101 Gegner in der Luft abgeschossen und sind einschliesslich Ihrer Erfolge im spanischen Bürgerkrieg 115 mal Sieger im Luftkampf gewesen.
>
> In Würdigung Ihres immer währenden heldenmütigen Einsatzes im Kampf um die Freiheit unseres Volkes und in Anerkennung Ihrer hohen Verdienste als Jagdflieger verleihe ich Ihnen als 1. Offizier der Deutschen Wehrmacht die höchste deutsche Tapferkeitsauszeichnung das Eichenlaub mit Schwertern in Brillanten zum Ritterkreuz des Eisernen Kreuzes.
>
> Mit meinem und des ganzen Deutschen Volkes Dank verbinde ich die besten Wünsche für Ihre Zukunft.

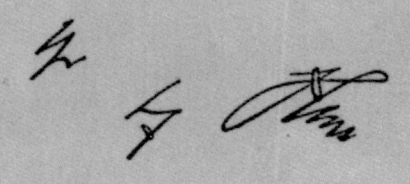

Adolf Hitler's letter on the occasion of the conferment of the "*Brillant.*"

German Fighter Ace Werner Mölders - An Illustrated Biography

Five days later he was promoted to *Oberst* (see certificate, left), and received the new decoration (from July 15, 1941) from Hitler on July 26, 1941, in FHQu. "Wolfsschanze."

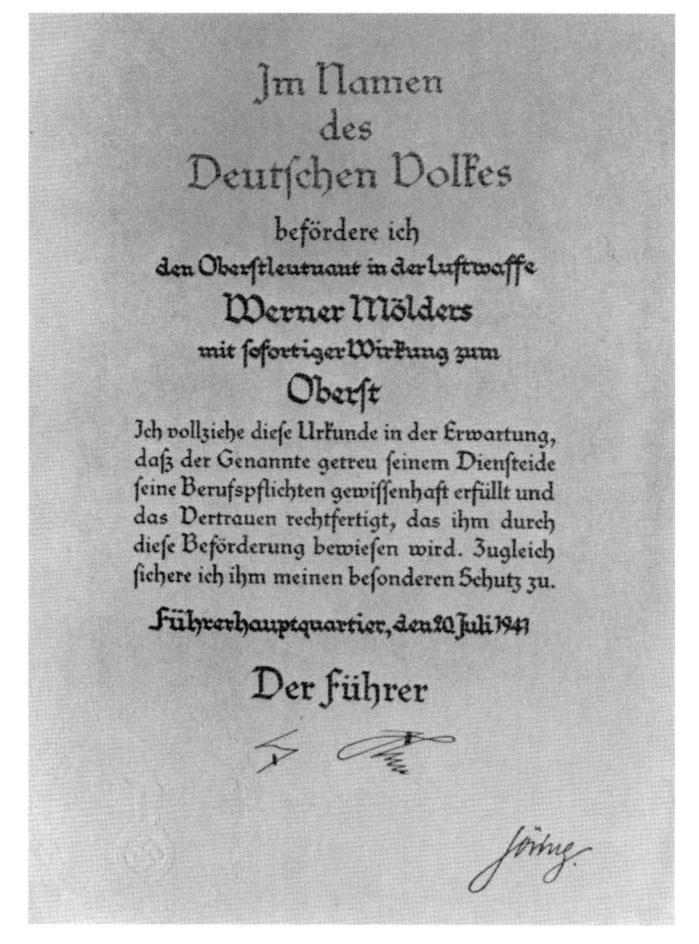

Oberst Werner Mölders, the first bearer of the *Brillant* in the German *Wehrmacht* (27 more conferments followed until the end of the war).

The valuable conferment certificate, photographed shortly before Hitler signed it. Description: National emblem, name, and text written in embossed gold on parchment, trimmed with sky blue maroquin frame with gold plating. Sky blue maroquin case with ivory velvet lining, handbound, National emblem, ornament, anc border edges in fire-gold, and the swastika inlaid with diamonds.

German Fighter Ace Werner Mölders - An Illustrated Biography

After the well-deserved honor Mölders received a grounding order, and had to give up his *Geschwader* for the appointment to *Inspekteur* of the fighter pilots. He flew from *Gruppe* to *Gruppe* and said goodbye to all of his men.

Both *Kommodore* and his *Feldwebel* Höfemeier found the departure difficult, as one can clearly see in the photograph. Left of Höfemeier is "Toni" Hafner.

Both successor *Kommodore* in one picture. Left Major Beckh, who took over the *Geschwader* after Mölders (missing since June 21, 1942, as *Kdre.* of JG 52 in Russia). From April 1942 until April 1944 Karlfried Nordmann led JG 51 with much success.

175

Inspekteur of the Fighter Pilots

The new *Inspekteur* flew and drove on the Eastern Front from unit to unit, in order to get an idea of all the problems, and to remedy the situation as quickly as possible.

The presentation of awards to deserving comrades was one of his pleasant duties. Here he puts the "*Dödel*" on Ltn. Erich Schmidt from III./JG 53, his former *Gruppe*, after his 30th air victory.

German Fighter Ace Werner Mölders - An Illustrated Biography

On July 27, 1941, his former flying instructor from München-Schleiflheim, Robert Olejnik, also received the *Ritterkreuz*. Mölders was happy for him.

In serious discussion about the situation with *Gen.Major* Hoffmann von Waldau, one of the most capable *Generale* of the *Luftwaffe*. Unfortunately, he suffered a pilot's death on May 17, 1943, as a passenger of a He 111 during bad weather in the Rhodopen (Bulgaria/Greece/Macedonia), coming upon a mountain that was entered in too low on the map.

German Fighter Ace Werner Mölders - An Illustrated Biography

Again and again Mölders was called into Göring's headquarters. The pictures show him in front of the special train of the *Oberbefehlhaber* of the *Luftwaffe* and with Göring, who wears his white *Reichsmarschall* uniform.

At the end of August 1941 Mussolini visited Hitler and Göring in their headquarters. Göring, almost hidden by the "Duce," introduced Germany's most successful fighter pilot.

German Fighter Ace Werner Mölders - An Illustrated Biography

Mölders was often present for official meetings, and as a guest in the "Karinhall," Göring's beautiful country house and place of residence in the Schorfheide, northeast of Berlin.

On September 14, 1941, Mölders married in the little city Falkenstein im Taunus. The snapshot shows him at a party on the eve of the wedding, where there was plenty of traditional broken glass.

Pfarrer Erich Klawitter, Mölders' father-like friend, blesses the bride and groom.

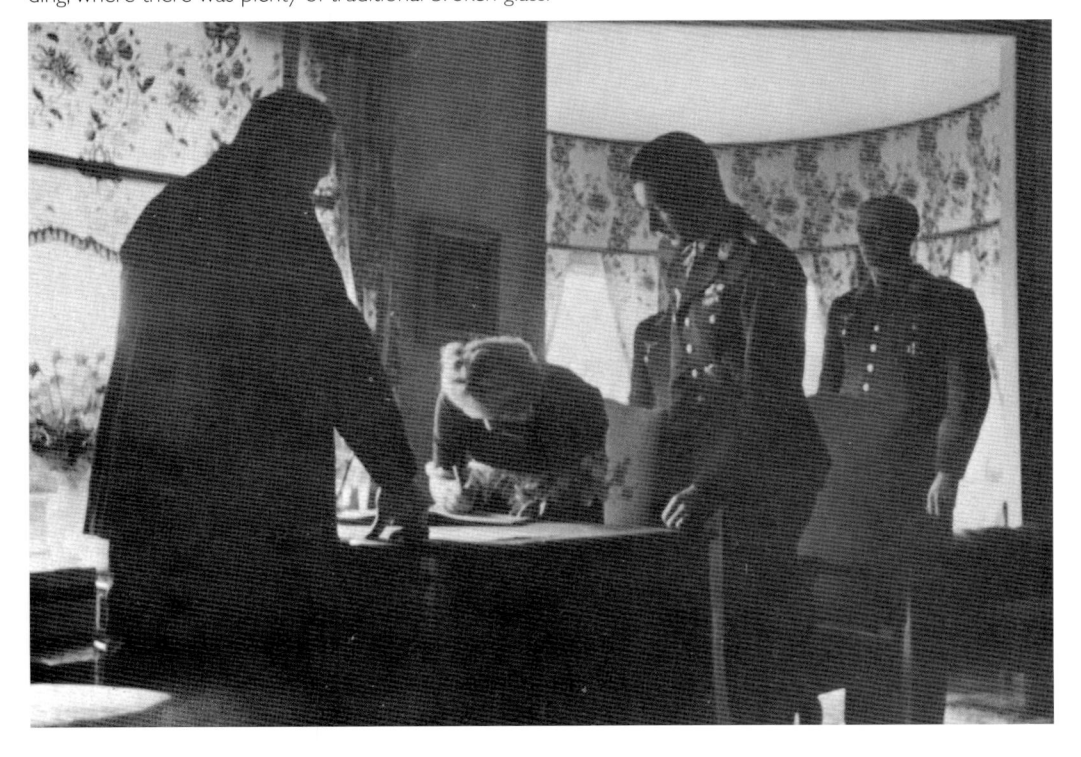

The bride's signature in front of the registrar. To the right of Mölders, his witness *Ltn*. Erwin Fleig, Left, and somewhat hidden, is witness *Obltn*. Hartmann Grasser.

German Fighter Ace Werner Mölders - An Illustrated Biography

After the church wedding; Luise and Werner Mölders with the witnesses and wedding guests.

181

The bride and groom with a few of the guests. Second from the left, Fritz von Forell, 7th from the left, his sister Annemarie Mölders, 9th from the left, *Mutti* Mölders, and to the far right, brother Hans Mölders.

After the wedding they go on their honeymoon for a few days in civilian clothes.

One of the many letters of congratulations that Mölders and his wife received when the wedding became known.

Keßelring
Generalfeldmarschall

Gefechtsstand, den 30. 9. 1941.

Lieber Mölders !

Verspätet bringe ich Ihnen meine Glück=
wünsche zu Ihrer Vermählung. Dass sie so spät
kommen, sind Sie diesmal alleine schuld.

Trotz Krieg und stärkster Inanspruchnahme
hoffe ich, dass Sie Ihrer jungen Frau etwas sein
können und wünsche, dass Sie in baldigem Frieden
eine schöne Ehe aufbauen können.

Mit der Bitte, mich Ihrer jungen Gattin zu
empfehlen, bin ich

stets Ihr

[signature]

At the end of September Mölders was on the Eastern Front again. Bottom left in conversation with *General d. Flg.* Loerzer, s.Zt. *Kommandierender General* of the II. *Flg.Korps*, and with Günther Lützow, *Kommodore* of JG 3.

Right: A visit with *Generallleutnant* Bodenschatz, representative of the *Luftwaffe* in the main headquarters. Bodenschatz was a successful fighter pilot in World War I with Manfred Frhr. von Richthofen.

German Fighter Ace Werner Mölders - An Illustrated Biography

Visit with allied *Fliegeroffiziere*.

Mölders also visited the volunteer pilots of the Spanish "EsquadrÛn AzˇI," whose *Kommandeur* Major Salas achieved the first shootdowns for the *Staffel* on October 4, 1941. The flag of the Bf 109-equipped Spanish *Jagdstaffel* shows the fallen Spanish Fliegeras Garcia Morato in the circle and the inscription, the war cry of the Spanish pilots during the civil war.

German Fighter Ace Werner Mölders - An Illustrated Biography

October 7, 1941, on *Platz* Karna: Mölders has just landed with a Bf 110. As *Bordschütze*, his *Adjutant*, Major Wenzel, flew with him, who here climbs out, somewhat tired. Cheerful welcoming of the old comrades from III./JG 51 under *Hptm*. Leppla (second from the right). Bottom right: In great spirits—"*Vati*" Mölders.

Wherever Mölders appeared, he was always surrounded by soldiers of all ranks. They all wanted to see their unforgettable former *Kommodore* and, when possible, speak a few words with him.

German Fighter Ace Werner Mölders - An Illustrated Biography

The three leading personalities of the German *Jagdwaffe*. From left to right: Karlfried Nordmann, *"Vati"* Werner Mölders, and "Franzl" Günther Lützow.

October 26, 1941, on the Krim. *Gen.Oberst* Löhr (second from left) in a discussion with a Turkish General (right). Between Löhr and Mölders, *Major* Handrick, *Kommodore* of JG 77. Handrick was an Olympic champion in the pentathlon in 1936 in Berlin.

It has become cold on the Krim! It is only bearable in the "officers' mess tent" of JG 77 in fur, or the warm bomber jackets. Next to Mölders is *Oberst Frhr*. von Houwald.

On the Krim: Mölders observes the *Jagd* and *Stukaeinheiten* under his command with *Fliegerleitoffiziere*, and directs the mission of the units from the distanced combat post.

German Fighter Ace Werner Mölders - An Illustrated Biography

German Fighter Ace Werner Mölders - An Illustrated Biography

With his *Adjutant*, Major Wenzel (right), and *Hptm*. Stormer on the Tscheplinka air field on the Krim.

German Fighter Ace Werner Mölders - An Illustrated Biography

Farewell to Werner Mölders

On November 17, 1941, *Gen.Oberst* Ernst Udet took his life. His suicide was covered up by the state, and was portrayed as a fatal accident during the testing of a new weapon! For his state funeral, arranged by Hitler (see the picture of the invitation), Mölders was also ordered to Berlin with other highly decorated *Luftwaffenoffizere*.

Der Reichsmarschall des Großdeutschen Reiches
und Oberbefehlshaber der Luftwaffe

beehrt sich,

zu dem am Freitag, dem 21. November 1941, 11 Uhr,

im Ehrensaal des Reichsluftfahrtministeriums in der Wilhelmstraße

stattfindenden

STAATSAKT ANLÄSSLICH DES STAATSBEGRÄBNISSES

des verewigten Generalluftzeugmeisters

Generaloberst Dr. h. c. ERNST UDET

einzuladen.

German Fighter Ace Werner Mölders - An Illustrated Biography

The last photographs of Werner Mölders. At the controls of his Fieseler "Storch," with which he then flew to Cherson.

Shortly before takeoff on November 21, 1941, on Cherson airfield (Ukraine) on the way to an He 111 of III./KG 27 "Boelcke," with which Mölders wanted to fly as a passenger to Berlin. Left, *Hptm*. Frhr. von Beust, right Major Wenzel, Mölders' Adjutant.

German Fighter Ace Werner Mölders - An Illustrated Biography

In conversation with bearer of the *Ritterkreuz Hptm*. Frhr. von Beust, the *Gruppenkommandeur* of III./KG 27...

...and with the crew members of the plane.

German Fighter Ace Werner Mölders - An Illustrated Biography

This photograph, taken secretly by an amateur photographer after the crash on November 22, 1941, is the only existing picture of the crash site in Breslau-Schöngarten. It is clear to see that the fuselage behind the cockpit broke off.

Commemorative plaque for *Oberst* Mölders that the property owner let have erected on the crash site in Breslau-Schöngarten.

Oberst Mölders †
Der Sieger in 115 Luftkämpfen tödlich verunglückt

Berlin, 23. Nov. Ein hartes Geschick hat es gefügt, daß die deutsche Luftwaffe wenige Tage nach dem Heimgange des Fliegerhelden aus dem Weltkrieg, Generaloberst U d e t, nun auch den kühnsten und besten aus den Reihen ihrer jungen Jagdflieger verlor: der Inspekteur der Jagdflieger, Oberst Werner M ö l d e r s, ist am 22. 11. auf einem Dienstflug mit einem Kurierflugzeug, das er selbst nicht steuerte, bei Breslau tödlich abgestürzt. Vom Feinde unbesiegt, fand der „Sieger in 115 Luftkämpfen" auf so tragische Weise den Fliegertod. Die Leistungen und Erfolge dieses von glühendem Kampfgeist beseelten erst 28jährigen Offiziers sind ohne Beispiel.

In Würdigung der einmaligen Verdienste des Obersten Mölders hat der Führer und Oberste Befehlshaber der Wehrmacht verfügt, daß das bisher von Mölders zu so gewaltigen Siegen geführte Jagdgeschwader in Zukunft seinen Namen trägt. Zugleich hat der Führer für Oberst Mölders ein Staatsbegräbnis angeordnet.

Die deutsche Luftwaffe verliert in Oberst Werner Mölders einen ihrer Besten. Das deutsche Volk aber und vor allem die deutsche Jugend verlieren in Werner Mölders einen ihrer großen Nationalhelden, an dem sie mit Verehrung, Bewunderung und Liebe hängen. Die ganze deutsche Nation steht tief erschüttert an der Bahre Werner Mölders, erfüllt von tiefer Trauer über den allzu frühen Tod, erfüllt aber auch vom Stolz auf diesen tapferen Sohn, dessen Ruhmestaten unvergänglich sind.

Als Angehöriger der Legion Condor ging Mölders 1938 nach Spanien, wo er im Kampf gegen den Bolschewismus mit 14 bestätigten Abschüssen der erfolgreichste Jagdflieger der Legion wurde. Mit dem Spanienkreuz in Gold mit Brillanten sowie mit der Medaille de la Campana und der Medaille Militar ausgezeichnet, kehrte er in die Heimat zurück.

Nach seinem 20. Abschuß im Kampf gegen England verlieh ihm der Führer im Mai 1940 das Ritterkreuz des Eisernen Kreuzes. Mit über 50 Gesamtabschüssen, davon 26 an der Westfront, stand er im Oktober 1940 an der Spitze der deutschen Jagdflieger. Als zweitem Offizier der deutschen Wehrmacht erhielt er aus Anlaß seines 40. Luftsieges am 22. September 1940 das Eichenlaub zum Ritterkreuz des Eisernen Kreuzes. Vom 22. Juni 1941, dem Beginn des Ostfeldzuges, bis zum 17. Juli schoß das Jagdgeschwader Mölders 500 Sowjetflugzeuge bei nur drei eigenen Verlusten ab und konnte damit insgesamt 1200 Luftsiege verzeichnen. Mölders selbst errang am Tage des Beginns des Entscheidungskampfes gegen den Bolschewismus seinen 72. Luftsieg. Am 23. Juni 1941 verlieh der Führer Major Mölders in Anerkennung seines erfolgreichen Einsatzes die Schwerter zum Ritterkreuz des Eisernen Kreuzes mit Eichenlaub, das ihm als zweitem Offizier der deutschen Wehrmacht verliehen wurde. Am 17. Juli erlegte er allein bei einem Luftkampf fünf Sowjetflugzeuge; damit hatte er im Verlaufe des Krieges insgesamt 101 Luftsiege erzielt. In Würdigung dieser ganz außerordentlichen Erfolge verlieh der Führer am 16. Juli 1941 Werner Mölders als erstem Offizier der deutschen Wehrmacht die höchste deutsche Tapferkeitsauszeichnung, das Eichenlaub zum Ritterkreuz des Eisernen Kreuzes mit Schwertern und Brillanten.

Phot.: Archiv.

„Eine strahlende Heldengestalt"
Rom und Madrid zum Tode von Oberst Mölders

Die gesamte römische Sonntagspresse bringt zum tragischen Tod von Oberst Mölders das tiefe Mitempfinden des italienischen Volkes zum Verlust „dieser strahlenden Heldengestalt", wie „Giornale d'Italia" schreibt, zum Ausdruck. „Der Held aller deutschen Lufthelden mußte einem ganz einfachen Unfall zum Opfer fallen", schreibt das Sonntagsblatt „Lavoro Fascista". Bei Freund und Feind sei der Ruhm von Oberst Mölders geradezu legendär gewesen. Nun sei der Held in Walhall eingezogen, ohne daß jemals sich jemand werde rühmen können, ihn besiegt zu haben.

*

Die Madrider Presse widmet dem verunglückten deutschen Fliegerhelden Oberst Mölders herzliche Nachrufe und veröffentlicht Bilder von ihm. „Die Falange", so schreibt „Arriba", „teilt den Schmerz des deutschen Volkes, denn Mölders setzte sein Leben als Flieger der Legion Condor auch für Spaniens Freiheit ein. Mölders war mit seinen 28 Jahren das heldische Symbol dieses großen Krieges, das kämpferische Vorbild eines neuen werdenden Europa und die personifizierte Stärke der deutschen Luftwaffe. Gott schien diesem Flieger, der seine ersten 14 Luftsiege am spanischen Himmel errang, das Leben als eine besondere Gnade über alle Gefahren hinweg erhalten zu wollen, aber Mölders mußte das tragische Geschick des spanischen Fliegerhelden Morato, des italienischen Italo Balbo und Udets teilen. Mölders verschaffte keinem feindlichen Flieger ein Siegesmal auf dem Steuer seiner Maschine. Deutschlands Fliegerelite verfügt noch über hundert andere Männer, die die Abschußziffern Mölders erreichen werden, doch Mölders wird immer der erste sein."

*

Bei dem tragischen Flugunfall des Obersten Mölders fanden auch die Besatzungsmitglieder, Flugzeugführer Oberleutnant K o l b e und der Bordmechaniker Feldwebel H o b I e, den Fliegertod. Der Adjutant von Oberst Mölders sowie der Bordfunker wurden verletzt.

One of the countless number of press reports on Mölders' death.

Ein Jagdgeschwader führt den Namen Mölders
Das ganze deutsche Volk trauert um den unvergleichlichen Fliegerhelden

dnb. Berlin, 22. November.

Ein hartes Geschick hat es gefügt, daß die deutsche Luftwaffe wenige Tage nach dem Heimgang des Fliegerhelden aus dem Weltkrieg, Generaloberst Udet, nun auch den Kühnsten und Besten aus den Reihen ihrer jungen Jagdflieger verlor:

Der Inspekteur der Jagdflieger, Oberst Werner Mölders, ist am 22. November auf einem Dienstflug mit einem Kurierflugzeug, das er selbst nicht steuerte, bei Breslau tödlich abgestürzt. Vom Feinde unbesiegt, fand der Sieger in 115 Luftkämpfen auf so tragische Weise den Fliegertod. Die Leistungen und Erfolge dieses von glühendem Kampfgeist beseelten, erst 28jährigen Offiziers sind ohne Beispiel. Am 15. Juli 1941 verlieh der Führer und Oberste Befehlshaber der Wehrmacht dem Commodore Oberst Mölders nach seinem 101. Luftsiege im Freiheitskampf des deutschen Volkes als erstem Soldaten der Wehrmacht die höchste Tapferkeitsauszeichnung: das Eichenlaub mit Schwertern und Brillanten zum Ritterkreuz des Eisernen Kreuzes.

In Würdigung der einmaligen Verdienste des Obersten Mölders hat der Führer und Oberste Befehlshaber der Wehrmacht verfügt, daß das bisher von Mölders zu gewaltigen Siegen geführte Jagdgeschwader in Zukunft seinen Namen trägt. Zugleich hat der Führer für Oberst Mölders ein Staatsbegräbnis angeordnet.

Er bleibt unvergessen
Drahtbericht unserer Berliner Schriftleitung
ms. Berlin, 23. November.

Wir können es noch nicht fassen —: Mölders, unser Mölders ist nicht mehr. Der beste Jagdflieger der Welt, der Held und Sieger in 115 Luftkämpfen (einschließlich Spanien), das Vorbild der Jugend, der einzigartige Offizier und Kamerad ist tot. Erschüttert stehen wir alle im Geist an der Bahre dieses wunderbaren Mannes, dem die höchste Tapferkeitsauszeichnung überhaupt, die die deutsche Wehrmacht zu vergeben hat, zuteil wurde.

Die militärische Laufbahn Werner Mölders, der ein so tragisches Ende gefunden hat, ist eine fast meteorhafte. Am 18. März 1913 in Gelsenkirchen als Sohn eines Studienrates, der 1915 als Reserveoffizier gefallen ist, geboren, besuchte er das Realgymnasium in Brandenburg, trat 1931 in das Infanterieregiment 2 ein und wurde 1934 zum Leutnant befördert. Im darauffolgenden Jahr trat er zu der wiedergegründeten Luftwaffe über und wurde 1936 zum Oberleutnant befördert. Mit der Legion Condor ging er zwei Jahre später nach Spanien, wo er innerhalb kurzer Zeit 14 bestätigte und einige unbestätigte Abschüsse erzielte. Als erfolgreichster Jagdflieger der Legion, ausgezeichnet mit dem Spanienkreuz in Gold und Brillanten, kehrte er in die Heimat zurück und wurde wegen hervorragender Leistungen außer der Reihe zum Hauptmann befördert.

Nachdem er vorübergehend zum Reichsluftfahrtministerium kommandiert war, erfolgte im März 1939 seine Ernennung zum Kommandeur einer Jagdgruppe. Nach seinem 20. Abschuß im Mai 1940 vom Führer mit dem Ritterkreuz zum Eisernen Kreuz ausgezeichnet und kurz darauf zum Major befördert, stand er im Oktober 1940 mit über 50 Gesamtabschüssen, davon 26 an der Westfront, an der Spitze aller deutschen Jagdflieger. Auch im Kampf gegen Großbritannien leistete Mölders Hervorragendes. Als erster Offizier der Luftwaffe erhielt er zu seinem 40. Luftsieg am 22. September 1940 das Eichenlaub. Einen Monat später waren es bereits 50 Abschüsse, am 25. Oktober wurde er zum Oberstleutnant befördert, nach seinem 72. Luftsieg erhielt er das Eichenlaub mit Schwertern zum Ritterkreuz. Neuer Einsatz: Kampf gegen die Bolschewisten. Wieder ist Mölders dabei, wieder ist er einer der Tapfersten. Am 16. Juli 1941 kommt eine Sondermeldung aus dem Führerhauptquartier: Mölders hat 101 feindliche Flugzeuge, außer seinen Spanienerfolgen, abgeschossen. Als erster Offizier der deutschen Wehrmacht erhält er das Eichenlaub mit Schwertern und Brillanten zum Ritterkreuz des Eisernen Kreuzes. Gleichzeitig wurde er, der 28-Jährige, zum Oberst befördert.

Mitten im Einsatz für den Sieg erreicht Mölders die Nachricht von dem tödlichen Unglücksfall des Generalobersten Udet. Nur für wenige Stunden eilt er nach Berlin, um an dem Staatsakt für diesen Mann teilzunehmen, der an maßgeblicher Stelle die Grundlagen auch für die Waffe geschaffen hatte, die Mölders in so hervorragender Weise führte und meisterte. Auf dem Flug nach Breslau erreichte ihn das widrige Geschick. Es war ihm nicht vergönnt, vor dem Feind zu fallen. Und das ist das Tragische an seinem Tod. Wir müssen Abschied nehmen von Werner Mölders. Sein Geist aber lebt weiter, nicht nur in der Luftwaffe, deren bester er war, sondern im ganzen deutschen Volk, dessen Jugend seinen Taten nacheifern wird.

The armband "Jagdgeschwader Mölders" that all members of JG 51 wore with pride until the end of the war.

German Fighter Ace Werner Mölders - An Illustrated Biography

24. November 1941 * Nr. 328 * Seite 2

Auf einem Dienstflug von der Ostfront nach Berlin ließ am 22. November beim Absturz einer Kuriermaschine, die er nicht selbst steuerte,

Oberst Werner Mölders

im Alter von 28 Jahren sein junges Leben für Führer und Vaterland.

Tief erschüttert und auf das schwerste getroffen von diesem harten Schicksalsschlag, steht das deutsche Volk an der Bahre seines größten Helden im Kampf um Deutschlands Freiheit und Ehre, in stolzer Trauer um den Offizier, der bis heute als einziger Soldat der deutschen Wehrmacht die höchste Tapferkeitsauszeichnung: das Eichenlaub mit Schwertern und Brillanten zum Ritterkreuz des Eisernen Kreuzes in Siegesbewußtsein und Bescheidenheit trug.

Ehrfurchtsvoll senken sich die Fahnen aller Waffenteile der deutschen Wehrmacht vor dem ruhmgekrönten jüngsten deutschen Obersten, dem ob seiner edlen Gesinnung und seines überragenden Heldentums die Herzen aller, der Vorgesetzten wie der Kameraden und Untergebenen, entgegenschlugen, der sich die Liebe und Bewunderung der begeisterten deutschen Jugend wie kein anderer in diesem Kriege erworben hatte.

Ergriffen nimmt mit mir die deutsche Luftwaffe nun Abschied von dem Tapfersten aus ihren Reihen, dem vorwärtsstürmenden Kämpfer, der in mehr als 1000 Luftschlachten stets Sieger blieb und in beispielhaftem Angriffsgeist 115 feindliche Flugzeuge vernichtete. Unbesiegt in allen Kämpfen ist der hervorragendste deutsche Flieger, der Offizier, der allen Vorbild war und immer bleiben wird, das Opfer eines tragischen Unfalles geworden.

Sieg und Ruhm ist an seinen Namen geheftet, der in der Geschichte dieses Freiheitskrieges und der deutschen Zukunft ewig fortleben wird.

Durch die enge Verbundenheit, die ich stets für diesen jungen Kameraden empfand, trifft mich sein Tod persönlich aufs tiefste.

Göring
Reichsmarschall des Großdeutschen Reiches und Oberbefehlshaber der Luftwaffe

---

Tagesbefehl des Reichsmarschalls Göring

„Seinem kühnen Angriffsgeist sollt ihr nacheifern"

Berlin, 24. November.

Reichsmarschall Göring hat zum Fliegertod des Obersten Mölders den nachstehenden Tagesbefehl an die Luftwaffe erlassen:

Soldaten der Luftwaffe!

Unser Oberst Mölders weilt nicht mehr unter uns. Eine unerforschliche Vorsehung hat es gewollt, daß der Sieger in 115 Luftkämpfen, der Offizier, der als einziger in der deutschen Wehrmacht das Eichenlaub mit Schwertern und Brillanten zum Ritterkreuz des Eisernen Kreuzes als höchste Tapferkeitsauszeichnung trug, das Opfer eines tragischen Flugzeugunfalles geworden ist.

In tiefer Erschütterung treten wir an die Bahre unseres Besten und Tapfersten. Unfaßbar ist uns allen, daß unser ruhmreichster Flieger nicht mehr in unseren Reihen steht. Wie ein strahlender Komet zog sein junges Heldenleben hellleuchtend als Beispiel unbesiegbaren Kampfeswillens und vorbildlicher Tapferkeit an uns vorüber. Siegreich auf allen Schlachtfeldern dieses Krieges um Deutschlands Ehre und Freiheit, hat ihn kein Feind überwältigen können. Nun ist er, der treueste Pflichterfüllung und höchste Einsatzbereitschaft verkörperte, in Walhall eingezogen.

Auf Befehl des Führers und Obersten Befehlshabers der Wehrmacht trägt sein siegreiches Geschwader nun seinen Namen. So wird er in der Luftwaffe wie in der Geschichte des deutschen Volkes bis in alle Ewigkeit fortleben. Sein Andenken soll uns stolze Tradition und stets Vorbild höchster militärischer Tugend sein. Seinem kühnen Angriffsgeist sollt ihr nacheifern, um so die Lücke zu schließen, die sein Tod in unsere Reihen gerissen hat. Darum vorwärts, Kameraden, zum Endsieg im Geist unseres unvergeßlichen Helden.

Göring,
Reichsmarschall des Großdeutschen Reiches und Oberbefehlshaber der Luftwaffe.

Obituary and *Tagesbefehl* of the *Reichsmarschall*, published in all of Germany's big newspapers and for the troops on the front.

November 24, 1941: *Oberst* Mölders was laid out in the chapel of a military hospital reserve in Breslau. To the right of him in a closed coffin, the pilot *Obltn.* Kolbe, and left flight mechanic *Ofw.* Hobbie, who died with him. Bearer of the *Ritterkreuz Oberstltn.* Ludwig Schulz, *Kdr.* of *Luftkriegsschule* 5, shows his respect and lays down the *Reichsmarschall's* wreath.

One day later the transport to Berlin followed. All of Breslau took part in it, and tens of thousands of people lined the last path of Mölders with genuine interest and sadness, which also went by the birthplace of his greatly admired fighter pilot, Frhr. Manfred von Richthofen.

Abschied von Oberst Mölders
Staatsakt in Anwesenheit des Führers

Berlin, 28. Nov. Die Trauer des deutschen Volkes um Werner Mölders fand ergreifenden Ausdruck in einem feierlichen Staatsakt im Ehrensaal des Reichsluftfahrtministeriums, bei dem Reichsmarschall Hermann Göring in Gegenwart des Führers die Persönlichkeit dieses hervorragendsten deutschen Offiziers in ihrer ganzen beispielhaften Größe würdigte, und in dem vom Führer angeordneten Staatsbegräbnis.

Alle in Berlin anwesenden führenden Männer des nationalsozialistischen Staates, höchste Offiziere der deutschen Wehrmacht, Reichsminister, Reichsleiter und Gauleiter, Staatssekretäre sowie Angehörige der diplomatischen Vertretungen der befreundeten Nationen erwiesen dem toten Helden die letzte Ehre. Mölders fand auf dem Invalidenfriedhof neben Manfred von Richthofen und Ernst Udet seine Ruhestatt.

In der elften Stunde zog in der Wilhelmstraße vor dem Reichsluftfahrtministerium ein Bataillon der Luftwaffe und eine Batterie eines Flakregiments zur Trauerparade auf. In Begleitung des Reichsmarschalls und des Generalfeldmarschalls Milch schritt der Führer die Trauerparade ab. Danach begab sich der Führer zum Ehrenmal, wo die Bahre von Oberst Mölders ruhte, bedeckt mit der Reichskriegsflagge, dem Stahlhelm und einem herrlichen Strauß roter Rosen.

Reichsmarschall Göring widmete Werner Mölders folgende Worte:

„Genau eine Woche ist es her, als unsere ruhmreichen Fahnen sich hier senkten. Da nahmen wir Abschied von einem Helden des großen Weltkrieges, schmerzlichen Abschied von meinem alten treuen Waffengefährten.

Und heute heißt es nun wieder Abschied nehmen. Noch waren wir im Bann des schweren Schicksalsschlages, der uns den siegreichen Helden des Weltkrieges nahm; da kam die schier unfaßbare Nachricht, daß du, mein junger Freund, von uns gegangen bist. Unfaßbar deshalb, weil dein ganzes Leben ein Sieg war, unfaßbar für uns, weil wir nicht glauben konnten, daß deine kraftvolle, lebensbejahende Gestalt nun nicht mehr deinen Kameraden voranleuchten soll.

Vom Feinde unbesiegt bist du gefallen. An keiner der vielen Fronten, an denen du gekämpft und dein Geschwader von Sieg zu Sieg geführt hast, konnte dich der Gegner überwinden! Doch das Schicksal, das stärker ist als wir alle, das für uns unerforschlich bleibt, hat dich nun auf so tragische Weise abberufen.

Tiefe Trauer bewegt uns. Und doch wird sie überstrahlt von der stolzen Freude und dem Glück, daß wir dich, du junger Held, besitzen und unser nennen durften. Leicht wird das Wort Held ausgesprochen, und nur wenigen ist es vergönnt, wahres Heldentum zu erringen. Und fast immer ist es so gewesen, daß die Gestalt des Helden im Bewußtsein des Volkes jugendlich erschien und auch jugendlich das Erdendasein verließ. Unsere alten Vorfahren versetzten solch unsterbliche Helden in ihrem Glauben an die ewige Kraft des Sieghaften dann in den hohen Himmelsdom und ließen sie weiter dort als Sterne uns erscheinen. Oder sie grüßten sie auf ihrer Fahrt nach Walhall im Bewußtsein, daß nur der Kämpfer, nur der starke Mann, daß nur der Held, der, das eigene Leben nicht achtend, für sein ganzes Volk Schicksal trug und Schicksal wurde, ewigen Ruhmes und höherer Ehre teilhaftig sein sollte als all die andere Menschheit.

Sie sind nie alt geworden, die ganz großen Helden. Strahlend in ihrer Jugendkraft blieben sie unvergeßlich in unserem Gedächtnis.

Und so bist auch du gewesen: unerschrocken, jung und strahlend, das Abbild des kühnen Kämpfers. Während all der harten Kämpfe bist du immer Sieger geblieben! Unsterblich sind deine Taten. Unsterblich bleibt dein Name. Du bist so vielen Vorbild gewesen und hast so viele deiner jungen Kameraden auch zu Helden erzogen. Ein stolzes Gefühl bewegt mich, daß diese deine Kameraden, alle ebenfalls harte Kämpfer und siegreiche Helden wie du, nun dir, unserem Kühnsten und Tapfersten, in getreuer Kameradschaft an deiner Bahre den letzten Ehrendienst erweisen. Sei gewiß, daß diese Kameradschaft nie erlöschen wird.

Und wie nahe du meinem Herzen gestanden hast, weißt du selbst, du junger, glücklicher, du sieggekrönter strahlender Mensch.

Als vorhin die unsterblichen Töne aus Richard Wagners „Götterdämmerung" erklangen, da verstummte auch unsere Trauer, da fühlten wir nur das Strahlende und Sieghafte. Unter diesen Klängen, unter diesen Fanfarenstößen bist du emporgestiegen. Jetzt wird uns die Kraft deines Heldentums von dort leuchten. Immer wird dein Vorbild vor uns stehen. Und so lange es deutsche Jungen gibt und so lange eine deutsche Luftwaffe die Heimat schirmt, wird nur einen Wunsch haben: ein Mölders zu werden!

Gewiß, für uns, deine Kameraden, und für deine Angehörigen ist es bitter und schwer, daß du jetzt von uns gehen mußt. Aber für dich selbst ist alles so schön gewesen, so groß, so glänzend. Und letzten Endes werden auch die Menschen, die dein Verlust am schwersten getroffen hat, trotz aller Trauer das Glück empfinden, dir nahegestanden zu haben, ob es die Mutter ist, die dich geboren hat, ob die Geschwister, die dich begleitet haben, oder die Frau, die dich jetzt hergeben muß, oder die Kameraden, die dich nun missen müssen. Ueber all dieser Trauer steht der unerhörte Glanz deines Sieges, der den deutschen Volk die Ueberzeugung gegeben hat, daß Deutschland nur siegen kann. Ein Volk, das solche Helden hat, ist zum Siege bestimmt!

Und wenn der Tag kommt, da der Sieg errungen ist, der Tag, an dem die Banner der Freiheit und des Sieges aufgezogen werden und das deutsche Volk dankbar seiner Helden gedenkt, dann wirst du unter ihnen an erster Stelle stehen, denn du hast dem deutschen Volk die Gestalt des jungen Siegfried zurückgegeben, des strahlenden Helden, der früh in der Schönheit und der Kraft des jugendlichen Sieges gehen mußte.

Und nun darf ich zum Abschied dir sagen, du junger Held: Fahr auf nach Walhall!"

Pressemeldung mit der Schilderung des Staatsakts (Staatsbegräbnis) vom 28. November 1941.

Press report with the account of the act of state (state funeral) on November 28, 1941.

German Fighter Ace Werner Mölders - An Illustrated Biography

The viewing in the honoring yard of the *Reichsluftfahrtministerium* in Berlin.

Mölders' medals of bravery (from the left): *Goldenes Flugzeugführerabzeichen mit Brillanten, Eichenlaub mit Schwertern und Brillanten zum Ritterkreuz des Eisernen Kreuzes,* the *Deutsches Spanien-Kreuz in Gold mit Schwertern und Brillanten;* below: the *Eichenlaub zum Ritterkreuz,* the *Eichenlaub mit Schwertern zum Ritterkreuz des Eisernen Kreuzes.*

German Fighter Ace Werner Mölders - An Illustrated Biography

The act of state on 28 November in the *Reichsluftfahrtministerium* on the Wilhelmstraße in Berlin. Shortly before 11:00 am Hitler arrived with his escort. Left of him *Reichsleiter* Martin Bormann, right *Gen.Feldmarschall* Milch.

Hitler salutes the dead fighter pilot for the funeral service in the *Reichsluftfahrtministerium*'s ceremony hall.

The *Reichsmarschall*'s salute on the bier. His highly decorated fighter pilot comrades are on sentry duty. Left row: Schalk, Lützow, Oesau, and Müncheberg; right row: Galland, Falk, Kaminski, and Nordmann (hidden).

German Fighter Ace Werner Mölders - An Illustrated Biography

Left: While *Unteroffiziere* of the *Luftwaffe* set the coffin on the gun carriage that will lead Mölders for his last ride through Berlin, Göring and Milch salute the dead with a raised *Marschallstab*. Right: The decoration pillow, carried by a young *Oberleutnant* of the fighter pilots.

The funeral procession march through the *Reich*'s capital. Next to the gun carriage, the *Unteroffiziere* and the eight bearers of the *Ritterkreuz* are on sentry duty. Behind them Göring, five more bearers of the *Ritterkreuz* of the fighter pilots, the entire generals of the *Wehrmacht*, and the heads of state in Berlin at the time.

German Fighter Ace Werner Mölders - An Illustrated Biography

Like in Breslau, tens of thousands of people said goodbye to Mölders in Berlin, and lined the streets for his last journey to the *Invalidenfriedhof*. He found his last place of rest close to Frhr. von Richthofen and *Gen.Oberst* Ernst Udet, buried a week before.

The last part of his passage, by the grave of the most successful fighter pilot of World War I, *Rittmeister* Manfred Frhr. von Richthofen, fallen on April 21, 1918, in France, and brought back to his home in November 1925.

An 8.8 cm *Flak-Batterie* of the *Luftwaffe* shoots the *Ehrensalut* in the zoological gardens. In the background, right, the "Siegessäule."

German Fighter Ace Werner Mölders - An Illustrated Biography

"I had a comrade!" - With drawn swords at the grave (from right to left): Galland, Falk, Kaminski, and Nordmann. To the far right behind the bearers of the *Ritterkreuz* and fighter pilot comrades are Gordon M. Gollob and Hannes Trautloft.

The last salute!

German Fighter Ace Werner Mölders - An Illustrated Biography

The grave in the *Invalidenfriedhof* in Berlin, which was situated in the eastern part of the city. After the war the powers in East Berlin did not take care of the traditional cemetery, which suffered much damage in 1945, and let it go to rack and ruin. In 1975 they completely dissolved the cemetery and leveled all the graves! They destroyed a treasure of Prussian-German history. Not only great soldiers rested in the cemetery, but also famous inventors, scientists, writers, diplomats, and businessmen. May the following saying on the tombstone of the *Invalidenhausdichter* Major Wallaiser remember all and these unforgotten former German men and women of honor:

The old lindens rustle,
the evening closes in.
In the magic darkness
the silent grove of heroes.
They once fought for Germany,
made Prussia great,
they rest here in peace
in the cool night of the earth.
Be thankful, you brave men!
That, which is temporal, passes!
May that, which you have created,
last forever.

AM SONNTAG, DEM 14. DEZEMBER 1941, 11 UHR

findet in der Stadthalle zu Ehren des verewigten Inspekteurs der Jagdflieger

OBERST WERNER MÖLDERS

eine Gedenkfeier statt.
Wir beehren uns, hierzu ergebenst einzuladen.

Brandenburg ⟨Havel⟩, den 6. Dezember 1941.

DR. SIEVERS
Oberbürgermeister

HEPPNER
Kreisleiter I. V.

Zum Tode unseres lieben, unvergeßlichen Werner sind uns unendlich viele wohltuende und ehrende Beweise rührender Teilnahme zugegangen. Von ganzem Herzen danken wir dafür. Es tut uns leid, daß wir nicht jedem Einzelnen persönlich antworten können. Wir bitten deshalb unseren tiefgefühlten Dank auf diesem Wege entgegenzunehmen.

Im Namen aller Angehörigen

Annemarie Mölders
geb. Riedel
Brandenburg/Havel

Luise Mölders
geb. Thurner
München

Commemoration ceremony in Mölders' hometown, Brandenburg/Havel, and the acknowledgement from the Mölders Family for the countless number of condolences they received from all over Germany and befriended foreign countries, especially Spain.

German Fighter Ace Werner Mölders - An Illustrated Biography

For the lasting memory of Werner Mölders: The outstanding PK-drawing by Wolf Willrich, drawn in 1940; the beautiful oil painting by the famous war painter, Erich Cleff, that was created in 1941, as well as a bronze bust from 1940. The name of the artist who created it is, unfortunately, no longer known.

German Fighter Ace Werner Mölders - An Illustrated Biography

Unforgotten - *Oberst* Werner Mölders

Frau Luise Mölders with her daughter Verena, whose birth Werner Mölders, unfortunately, did not experience.

Verena, the daughter of Werner and Luise Mölders, eighteen years later.

German Fighter Ace Werner Mölders - An Illustrated Biography

April 13, 1958, in Bath/Maine (USA): The launching of the *Lenkwaffenzerstörer* "Mölders" from the shipyard "Bath Iron Works." It is the first time in the history of the German marines that a battleship was named after a well-deserving *Luftwaffenoffizer*.

The "Holiday Inn," where the evening festivities took place, greets its guests!

German Fighter Ace Werner Mölders - An Illustrated Biography

The launching was planned by Mölders' mother, and all family members were invited. From left to right: *Frau* Marlis with her husband Victor (younger brother), Annemarie Mölders (sister), Anna-Maria Mölders (the mother, 79 years old), and Hans Mölders (older brother) with his wife, Eleonore.

The *Lenkwaffenzerstörer* D 186 "Mölders" of the *Bundesmarine* in its element.

The former *Kommodore* of JG 51 "Mölders" for a visit on the *Zerstörer*. From the right: *Gen. Ltn. a.D.* Theo Osterkamp, Fritz Losigkeit, the *Kommandant*, *Fregattenkapitän* Mann, and Dr. Heinz Lange.

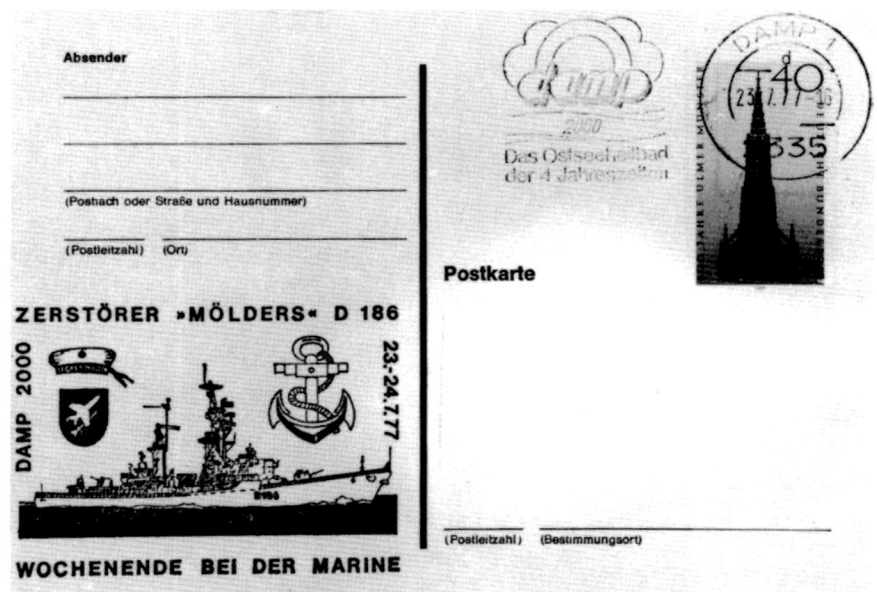

The *Zerstörer* "Mölders" goes on "great trips" again and again, and carries the name of this outstanding pilot, soldier, and man in all the world. Here a few philatelist records from these trips.

German Fighter Ace Werner Mölders - An Illustrated Biography

On November 9, 1972, Mölders was newly honored by the *Bundeswehr*. The *Luftwaffen-Kaserne* of II. *Abteilung* of the *Fernmelde-regiment* 34 in Visselhövede (in the Lüneburger Heide) received the name "Mölders-*Kaserne*."

On the fifth anniversary of II./*FmRgt*. 34 and the "*Tag der offenen Tür*" on September 18, 1976, this postal stationary was brought out.

2. LUFTWAFFEN DIVISION

Der Kommandeur der 2. Luftwaffendivision
Brigadegeneral Werner Schmitz

bittet

zu einem militärischen Appell anläßlich der feierlichen Verleihung des Traditionsnamens

„Mölders"

an das Jagdgeschwader 74 und Benennung der Kaserne Grünau in

„Wilhelm-Frankl-Kaserne"

mit anschließendem Empfang am 22. November 1973

nach

Neuburg an der Donau.

The greatest distinction after the war was given to the unforgettable fighter pilot on November 22, 1973, the 32nd anniversary of his death. In the form of a ceremonial line-up the name "Mölders" was given to *Jagdgeschwader* 74, stationed in Neuburg a. d. Donau. Mölders is the only soldier whose name is represented in two sections of the armed forces, the Marines and *Luftwaffe*.

German Fighter Ace Werner Mölders - An Illustrated Biography

November 22, 1973, at the Zell base in Neuburg/Donau: The *Inspekteur* of the *Bundesluftwaffe*, *Gen.Ltn.* (now a.D.) Günther Rall, one of the best fighter pilots of World War II, greets *Frau* Petzolt-Mölders. Left, Hans Mölders with his wife, and right Oberst Erlemann, at this time *Kommodore* of the JG 74.

Gen.Ltn. Rall and the *Kommodore*, *Oberst* Erlemann, during an inspection of the *Geschwader* lined up in parade formation for the ceremonial name-giving.

German Fighter Ace Werner Mölders - An Illustrated Biography

Dr. Heinz Lange, *Major a.D.* and bearer of the *Ritterkreuz zum Eisernen Kreuz*, last *Kommodore* of JG 51 "Mölders," during a ceremonial address.

View of the VIP lounge during the ceremony. Left in a light coat, and with a fur cap, the unforgotten Senior of the German fighter pilots, *Gen.Ltn. a.D.* "Uncle Theo" Osterkamp, who, despite his old age, took part in the honoring of his former student and successor as *Geschwaderkommodore*.

German Fighter Ace Werner Mölders - An Illustrated Biography

The new armband "*Geschwader* Mölders," and the band for the troop flag.

Generalleutnant Rall fastens the armband on the *Kommodore* of the *Geschwader*, *Oberst* Erlemann.

German Fighter Ace Werner Mölders - An Illustrated Biography

The new name of the *Geschwader*: *Jagdgeschwader* 74 "Mölders."

Right: *Generalleutnant* Rall (left) and *Oberst* Erlemann after the ceremonial line-up in front of a Bf 109-E with the *Geschwader*'s coat of arms of the former JG 51 "Mölders" painted on.

The famous Messerschmitt-*Jagdflugzeug* was a loan from the German Museum in München for the ceremony. The plane can be visited in the aeronautics section of the museum.

German Fighter Ace Werner Mölders - An Illustrated Biography

Left: During the same time as the conferment of the name, the *Geschwader*'s accommodations, the *Kaserne* Grünau, was named "Wilhelm-Frankl-Kaserne." Frankl was born on December 20, 1893, as the son of Jewish parents in Hamburg. In World War I he achieved a total of 19 air victories, and received the "pour le mérite" medal. On April 8, 1917, this outstanding fighter pilot, who was kept quiet about in the Third Reich as *Ltn. d. Res.* by Vitry-Sall (France), fell in air combat. Right: The accommodation area of *Jagdgeschwader* 74 "Mölders" is the first Kaserne of the *Bundeswehr* that is named after a German soldier of Jewish descent.

The memorial for Wilhelm Frankl in the yard of the *Fliegerkaserne* in Neuburg/Donau. After the unveiling of the memorial (from the left): *Oberst* Erlemann, Werner Nachmann, Chairman of the *Zentralrat* of Jews in Germany, *Gen.Ltn.* Rall, and *Gen.Ltn.* a.D. Theodor Osterkamp (Ü January 2, 1975).

219

German Fighter Ace Werner Mölders - An Illustrated Biography

Immortalized on letters and cards, the tidings of the name-giving to JG 74 is known to the entire world by the German Federal Post Office.

The members of *Jagdgeschwader* 74 "M" proudly wear the armband "*Geschwader* Mölders."

The new *Geschwader*'s coat of arms is painted on the modern fighter planes of JG 74 "M," here a McDonnell Douglas F-4 "Phantom."

German Fighter Ace Werner Mölders - An Illustrated Biography

In March 1975 a Tradition Room was opened in the "Wilhelm-Frankl-Kaserne." *Hptm*. Maison (right), who rendered outstanding services to the preservation of tradition, hands *Kommodore Oberst* Erlemann the model of a Fokker E-III *Jagdeinsitzer* from World War I.

In September 1979 the Tradition Room in the *Kaserne* was transferred and substantially enlarged. View of the exhibit in the new Tradition Room that is dedicated to the *Jagdflieger-As* of World War I, Wilhelm Frankl, and his comrades.

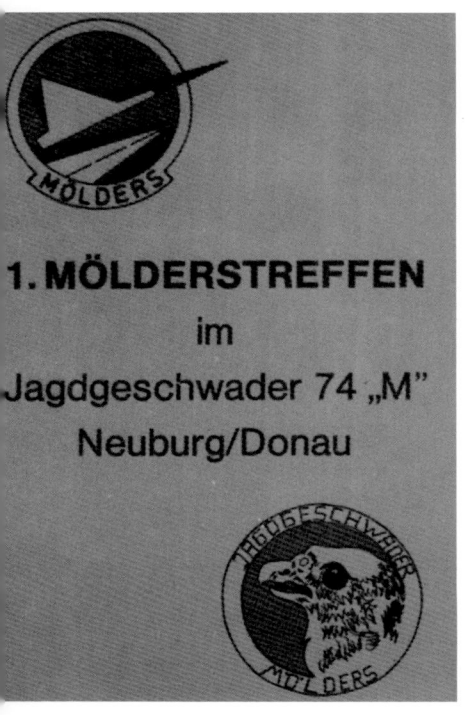

German Fighter Ace Werner Mölders - An Illustrated Biography

Left: From May 29-June 1, 1975, the first Mölders meeting took place in the "Wilhelm-Frankl-*Kaserne*" that united many old and new "*Mölderianer*," and ran a very harmonious course. Right: Guests Verena Buchanan, W. Mölders' daughter, and Erwin Fleig, the former *Kampfgefährte* and *Katschmarek* of her father.

Left: In a cheerful circle: *Oberst* Schmitz (left), the successor of *Oberst* Erlemann (right) as *Kommodore* of the *Geschwader*, *Frau* Petzolt-Mölders, and Dr. Heinz Lange. Right: Dr. Lange and *Hptm*. Maison during the unveiling of Mölders' portrait by Lothar Busse, a former pilot of JG 51 "Mölders." This gift from L. Busse hangs in the officers' mess of the "Wilhelm-Frankl-*Kaserne*."

German Fighter Ace Werner Mölders - An Illustrated Biography

JG 74 "M" received guests from the entire world, and the visitors viewed the outstandingly furnished Tradition Room of the *Geschwader*. Here a group of visitors of foreign *Stabsoffiziere* of the *Führungsakademie der Bundeswehr* in Hamburg with Oberst Schmitz (left) and *Hptm*. Maison (center).

Victor Mölders, Werner's younger brother, was one of the gladly seen guests. Here, with *Major* Riechmann, *Hauptmann* Maison, and Major Menge (from left to right), who also rendered outstanding services to the naming and keeping of tradition of *Geschwader* "Mölders."

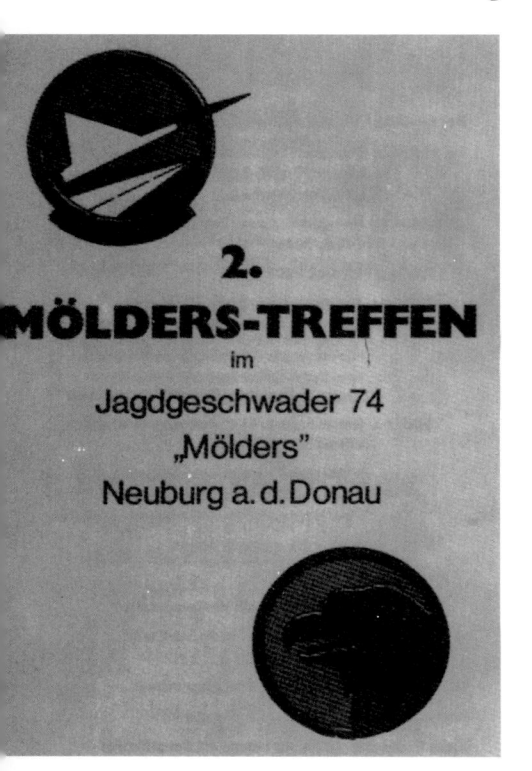

German Fighter Ace Werner Mölders - An Illustrated Biography

Left: The invitation to the second Mölders meeting from May 19-21, 1977, that was a huge success. Right: Two snapshots of this successful *Geschwader* meeting; *Frau* Petzolt-Mölders (right) in a conversation with Dr. Lange, the last *Kommodore* of JG 51 "Mölders," and his wife.

The beaming daughter of Werner Mölders, *Frau* Verena Buchanan, with her husband.

Top left: JG 74 "M" keeps in close contact with members of the former JG 51 "Mölders," especially with Werner Mölders' living widow in *Oberbayern*, *Frau* Luise Petzolt-Mölders. Here during a visit with the then *Kommodore Oberst* (now *Brigadegeneral*) Walter Schmitz.

Bottom right: For the third time since the naming, the former members of JG 51 with their families and the *Aktiven* of JG 54 "M" met from September 28-30, 1979, in Neuburg/Donau for the huge "Mölders meeting." The highlights were the opening and presentation of the new Tradition Room on the "*Tag der offenen Tür*," the "big tattoo," on the occasion of the 20 year existence of the *Bundeswehrgarnison* in Neuburg/Donau, as well as the unveiling of the "Mölders commemorative plaque" on the Base Zell.

German Fighter Ace Werner Mölders - An Illustrated Biography

View of the new Tradition Room with exhibits that remember Werner Mölders and his men of JG 51.

Among the souvenirs was the *Jagdflieger-Abschuflstock* of *Hptm*. Mölders for his achieved 23 air victories from September 1, 1939-June 3, 1940.

The model of the *Unglücks*-He 111 of KG 27 "Boelke," *Kennung* 1 G + TH, in which Mölders died on November 22, 1941.

In a display case, the original case with the conferment certificate for the *Eichenlaub mit Schwertern zum Ritterkreuz des Eisernen Kreuzes*.

German Fighter Ace Werner Mölders - An Illustrated Biography

September 28, 1979: The *Kommodore* of JG 74 "M," *Oberst* Michael Estendorfer (left), and *Major* Hartmut Hochgesang at the unveiling of the "Mölders-*Gedenktafel*," that was put up at the *Geschwader's* combat post on the Base Zell.

The bronze plaque for the name-giver of *Jagdgeschwader* 74 of the *Bundesluftwaffe*.

German Fighter Ace Werner Mölders - An Illustrated Biography

Unforgotten - *Oberst* Werner Mölders.

German Fighter Ace Werner Mölders - An Illustrated Biography

Sources

Printed Sources:
Fritz von Forell: "Mölders and his Men"
Fritz von Forell: "Mölders - Man and Pilot" (A Life Portrait)
Fritz von Forell: "Werner Mölders - Flight to the Sun" (The History of the Great Fighter Pilot)
Galland, Adolf: "The First and the Last"
Obermaier, Ernst: "The Bearers of the *Ritterkreuz* of the Luftwaffe 1939-1945. Vol. I - Figher Pilots"
Osterkamp, Theo: "Through the Highs and the Lows Hunts a Heart"
Ries, Karl/Ring, Hans: "Legion Condor 1936-1939" (An illustrated documentation)
Rübell, Günther: "Crosses - in Heaven - as on Earth"
Witetschek, Helmut: Quarterly Issue for Contemporary History, Issue 1 - January 1968
Diverse Editions of the "*Jägerblatt,*" the official organ of the "*Gemeinschaft der Jagdflieger E.V.*"
"*Der Völkische Beobachter,*" Years 1940 and 1941

Documents and photographs provided by:
Aders, Gebhard
Ing. (Grad.) Barbas, Bernd
Bardorf, Reinhold
Beckh, Friedrich
Fetzer, Rudolf
Fleig, Erwin
Held, Werner
Dipl. Ing. Architekt Hintze, Otto
Jagdgeschwader 74 "M"
Oberst a.D. Janke, Johannes
Joss, Karl
Dr. jur. Lange, Heinz
Lorant, Jean-Yves
Hptm. a.D. Maison, Ernst
Malo, Ernst
Mölders, Victor
Obermaier, Ernst
Olejnik, Rober
Petrick, Peter
Petzold-Mölders, Luise
Bauing. Ries, Karl
Ring, Hans
Schäfter, Wolf
Scherer, Rainer
Schnatmann, Gerhard
Weiler, Anton

A heartfelt thanks to all of those who offered their help. A special thanks from the authors goes to *Frau* Petzolt-Mölders, Werner Mölders' widow, for her husband's documents, like the service record book, flight book, certificates, letters, etc., for copying and printing, as well as her valuable collaboration on this work.

Thanks especially goes to *Herr Hptm. d.BW.a.D.* Ernst Maison, the *Traditionsoffizier* of JG 74 "M" for many years, for his support that contributed to the success of this book.

We would like to thank *Herr* Erwin Fleig for his foreword and the reviewing of this book, as well as the publishing house director Wolfgang Schilling for the publishing and setup of this work.

We would also like to thank *Herr* Victor Mölders and *Oberst a.D., Dipl. rer. pol.* Hans Gottfried Schulz for the reviewing of the manuscript and valuable advice.

Last, but not least we want to thank our mutual friend Hans Ring, who provided a valuable contribution to the development of this book with the review and revision of the manuscript.

München and Ransbach-Baumback, in October 1982
The Authors

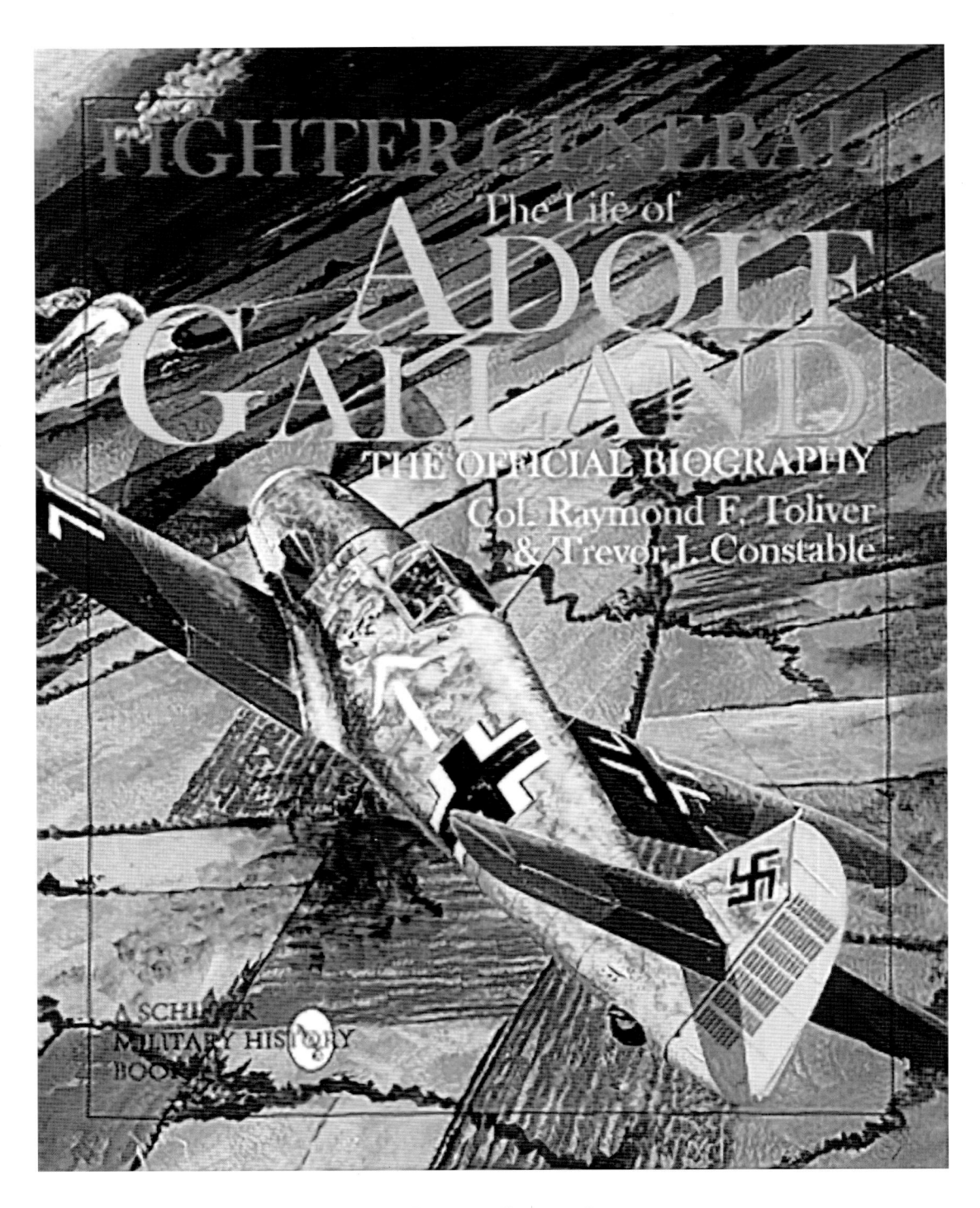

Fighter General
The Life of Adolf Galland
The Official Biography

Col. Raymond F. Toliver, USAF *(Ret.)* & Trevor J. Constable

Fighter General recounts the career of an outstanding combat leader torn from the fighter cockpit to defend his country – and sometimes his own pilots – in the bizarre bureaucracy of the Luftwaffe High Command. Here in this official biography is real-life adventure to shame the wildest fiction.
Size: 8 1/2" x 11"   ■   over 140 b/w photographs, 10 color aircraft profiles   ■   216 pp.
ISBN: 0-7643-0678-2   ■   hard cover   ■ $45.00